Maintenance of
Microorganisms

Maintenance of Microorganisms

A Manual of Laboratory Methods

Edited by
B. E. KIRSOP

National Collection of Yeast Cultures
Food Research Institute
Norwich, UK

and

J. J. S. SNELL

Division of Microbiological Reagents
 and Quality Control
Central Public Health Laboratory
London, UK

ACADEMIC PRESS, INC.
Harcourt Brace Jovanovich, Publishers
London   Orlando   San Diego
New York   Austin   Montreal   Sydney
Tokyo   Toronto

ACADEMIC PRESS INC. (LONDON) LTD
24/28 Oval Road, London NW1 7DX

United States Edition published by
ACADEMIC PRESS INC.
(Harcourt Brace Jovanovich, Inc.)
Orlando, Florida 32887

British Library Cataloguing in Publication Data

Maintenance of microorganisms.
  1. Microbiology—Culture and culture media
  2. Microbiology
  I. Kirsop, B. E.   II. Snell, J. J. S.

ISBN 0–12–410350–2

PRINTED IN THE UNITED STATES OF AMERICA

85 86 87 88   9 8 7 6 5 4 3

Contributors

I. J. Bousfield *National Collection of Industrial and Marine Bacteria, Torry Research Station, PO Box 31, 135 Abbey Road, Aberdeen, UK*

J. P. Cann *Culture Centre of Algae and Protozoa, Institute of Terrestrial Ecology, 36 Storey's Way, Cambridge, UK*

C. R. Contopoulous *Donner Laboratory, University of California at Berkeley, Berkeley, California 94720, USA*

H. Hippe *Deutsche Sammlung von Mikroorganismen, Gessellschaft für Biotechnologische Forschung mbH, Grisebachstrasse 8, D-3400, West Germany*

C. S. Impey *Food Research Institute, Colney Lane, Norwich NR4 7UA, UK*

E. James *Winches Farm Field Station, London School of Hygiene and Tropical Medicine, 395 Hatfield Road, St Albans, Herts, UK*

D. Jones *Department of Microbiology, School of Medical Science, University of Leicester, University Road, Leicester LE1 7RH, UK*

B. E. Kirsop *National Collection of Yeast Cultures, Food Research Institute, Colney Lane, Norwich NR4 7UA, UK*

E. A. Leeson *Culture Centre of Algae and Protozoa, Institute of Terrestrial Ecology, 36 Storey's Way, Cambridge, UK*

G. J. Morris *Culture Centre of Algae and Protozoa, Institute of Terrestrial Ecology, 36 Storey's Way, Cambridge, UK*

P. A. Pell *Department of Microbiology, School of Medical Science, University of Leicester, University Road, Leicester LE1 7RH, UK*

B. A. Phillips *National Collection of Dairy Organisms, National Institute for Research in Dairying, Reading RG2 9AT, UK*

R. H. Rudge *National Collection of Type Cultures, Central Public Health Laboratory, Colindale Avenue, London NW9 5HT, UK*

D. Smith *Culture Collection, Commonwealth Mycological Institute, Ferry Lane, Kew, Surrey TW9 3AF, UK*

P. H. A. Sneath *Department of Microbiology, School of Medical Science, University of Leicester, University Road, Leicester LE1 7RH, UK*

J. J. S. Snell *Division of Microbiological Reagents and Quality Control, Central Public Health Laboratory, Colindale Avenue, London NW9 5HT, UK*

S. A. Waitkins *Leptospira Reference Laboratory, Public Health Laboratory, County Hospital, Hereford HR1 2ER, UK*

Preface

Microbiology has always depended upon the maintenance of microorganisms to provide viable and stable cultures. The increasing use of microorganisms for industrial and other purposes compounds this need and makes effective culture maintenance a top priority. The UK Federation for Culture Collections, recognizing the demand for information on the subject, runs training courses to transfer the expertise developed in the national culture collections to those responsible for culture maintenance in teaching, research, medical and industrial laboratories. The courses are oversubscribed and this has suggested that the demand is not being met by courses alone. This book, which is based largely on material included in the training courses, enables dissemination of this information to a wider audience.

Information on culture maintenance is sparse and scattered widely through the literature; it is also uneven in its coverage for different groups of microorganisms. There are, therefore, substantial difficulties in finding reliable and up-to-date information. This book brings together in one volume methods that have been developed and used successfully by those experienced in culture maintenance. The manual includes chapters on bacteria (of general, medical and industrial importance; of anaerobes, methanogens and leptospira), fungi (including yeasts), algae and protozoa. In response to the demands of biotechnology, considerable effort is being directed towards the development of preservation methods for animal and plant cells, and chapters covering these topics will no doubt appear in future editions of this manual.

We are indebted to the authors of the sections of the book who have been enthusiastic in their willingness to share their experience. They will surely welcome feedback from microbiologists who will take the methods described in this volume, use them (possibly on different microorganisms) and develop them for their own purposes. Thanks are also due to the many colleagues who helped with typing and proof-reading and to Kevin Painting who prepared the index.

Two of the objectives of the UKFCC are "to promote communication between culture collections and their users" and "to promote the development of interest and research in preservation". The editors hope this book will further these objectives to the benefit of microbiology.

November 1983   B. E. Kirsop
Norwich and London   J. J. S. Snell

Contents

1

Introduction

J. J. S. SNELL
*Division of Microbiological Reagents
and Quality Control
Central Public Health Laboratory
London, UK*
and
B. E. KIRSOP
*National Collection of Yeast Cultures
Food Research Institute
Norwich, UK*

All microbiologists need to preserve microorganisms. This need applies equally to university, school, technical college, hospital, industrial, veterinary or agricultural laboratories. Just as chemistry or biochemistry would cease without a supply of pure chemicals, so microbiology is dependent on the availability of pure, stable cultures. Whereas, however, most chemicals are easily stored, microbial cultures are extremely vulnerable and may become contaminated, undergo change or die unless technical expertise is available to prevent this happening.

Microbiologists need to preserve cultures for a variety of reasons: teaching laboratories need to maintain a library of strains that give typical reactions; industrial laboratories need to maintain production strains or large numbers of strains for use in screening programmes; taxonomists must maintain large numbers of strains for comparative studies; and research laboratories require pure cultures for a multiplicity of purposes. Often cultures are irreplaceable and their loss may be serious; others may be re-isolated or replaced by one of the service culture collections. In either case, time, information, or money is lost, but this could often be avoided by an effective system for culture maintenance.

Despite the fundamental importance of a reliable supply of pure and stable cultures, culture maintenance is often afforded low priority and is frequently carried out with inadequate staffing levels and equipment. Moreover, industrial, university, or institute

MAINTENANCE OF MICROORGANISMS
ISBN 0 12 410350 2

collections are housed in laboratories where culture maintenance is not a primary function and time is seldom available to experiment with different preservation methods. The service culture collections, however, have a full commitment to the preservation of cultures and the opportunity to experiment with preservation techniques and their accumulated expertise is of great value to others.

Information on preservation methods is thinly distributed throughout the literature and often is not readily available; this manual assembles in one volume a collection of reliable preservation methods. Guidance is provided on the choice of suitable methods for particular circumstances. Technical details of methods are presented and their reliability with regard to expected survival levels and stability is assessed. Information is given on the expected shelf life of micro-organisms, the relative cost of different methods, sources of materials and equipment, and pertinent literature references.

It must be stressed that there is no universal method for successful preservation of all microorganisms. Taxonomic groups of microorganisms, and even strains within a species, will vary in their response to different methods of preservation. The methods presented in this manual will generally prove successful, but occasions will arise when further investigation will be needed to achieve complete success.

The manual is practical in nature and aims to provide details of methods which in the experience of the authors have been found to work well. Some chapters deal primarily with a specific method; others cover a number of methods found suitable for a particular group of microorganisms. It is possible, however, that methods described for one group of organism will prove useful for others. Repetition of detail has occurred in a number of chapters, but as the manual will be used primarily as a practical book this has been deliberately retained to avoid frequent reference to other chapters for essential methodology. The names and addresses of suppliers of suitable material (cited in the text) are provided in Appendix III. In general, the theoretical aspects of the effects of preservation methods on cell survival are not considered in detail, but readers with an interest in the subject are referred particularly to chapters 13 and 14 and to the references supplied at the end of the manual.

The service culture collections have accumulated much knowledge on the preservation and properties of the strains they maintain and can provide invaluable advice and information. The range of services provided by these collections is not always fully recognized and Chapter 3 is devoted to a description of them. Names and addresses of some major service collections are listed in Appendix II; those of all other collections may be obtained by reference to The World Directory of Collections of Micro-organisms (McGowan and Skerman, 1982).

Although the methods described in this manual aim to ensure the survival of microorganisms in an unchanged condition, quality control of the preserved cultures is needed to provide evidence of viability, purity and stability of characters, all of which must be satisfactory for the strain to be useful. The kind of quality control carried out will vary and is best determined by each laboratory. To ensure that maximum use is made of culture collections proper records must be maintained. Lists of the information that should form the basis of an effective documentation system are given in Appendix I (see also Chapter 14).

Failure to establish an effective programme of culture maintenance often leads to the death, contamination, genetic change or misplacement of important laboratory strains. To avoid the loss of time and increased expense that these difficulties incur, the preservation of cultures should be afforded the priority that its central position in microbiology demands.

2

Service Collections: their Functions

B. E. KIRSOP
National Collection of Yeast Cultures
Food Research Institute
Norwich, UK

The main function of service culture collections is to act as depositories for all kinds of microorganisms that are of past, present or potential importance, so that resource and information centres for the general support of microbiology are established. The basic function of culture collections, therefore, is the collection, maintenance and supply of cultures.

The organisms collected reflect the interests of the collection and are obtained in one of several ways. Following searches through the scientific literature direct application may be made to authors to deposit strains; alternatively microbiologists may themselves make approaches to appropriate collections. Again, collections themselves may deposit cultures that have formed part of their own taxonomic or identification activities. Deposited cultures are checked for purity and authenticity before accession by the collection.

The selection of maintenance methods that produce maximum survival levels and strain stability is of fundamental importance, since strain drift in stocks maintained in service collections is unacceptable. Assessments of survival are made immediately after processing and at intervals during storage, so that fall-off in viability is monitored and an effective maintenance programme established. In order to assure a reliable service for the supply of cultures, quality control measures are carried out routinely. In addition, appropriate stock-holding levels are determined and administrative arrangements are established for handling orders and invoicing customers; postal regulations are followed for the despatch of cultures at home and overseas.

These activities, common to all service culture collections, generate a quantity of essential information that must be systematically recorded. Many collections are now using computers for this

MAINTENANCE OF MICROORGANISMS
ISBN 0 12 410350 2

purpose. They are ideally suited to the storage and management of both customer records and stock supplies and greatly streamline the business administration of collections. Again, after appropriate coding, strain data may be stored in a computer, allowing rapid searching and retrieval of scientific information and establishing a data base that may be used subsequently for computer identification. Computers are being increasingly used for the storage of literature references and are equally appropriate for the preparation of catalogues. An up-to-date catalogue is an essential requirement for the effective functioning of a service collection and, in the past, its preparation was an onerous and lengthy job. The transfer of catalogue data to a computer, however, allows ready updating for future editions and provides copy that can be printed directly from disk or tape, thus eliminating the need for conventional proof-reading.

In addition to the three basic activities (the collection, maintenance and supply of cultures, and the record-keeping associated with them) culture collections provide a number of other services. As a result of the expertise developed by collection staff and the close contacts established with microbiologists using their strains, collections become information centres for all matters relating to the microorganisms they hold. In addition to supplying answers to written and telephoned enquiries, staff are called upon to lecture, help on courses and contribute to scientific journals and books. Much of the expertise existing in the collections is of a taxonomic nature and most service collections are able to provide an identification service. This is particularly valuable for microbiologists, since identification may be lengthy, expensive, demand sophisticated techniques, and yet only be required intermittently.

A further service of particular interest to industry is the provision of "safe-deposit" facilities. Several collections are able to maintain important strains under optimum conditions, while providing access only to the depositor. This enables industry to depend on a reliable back-up for the preservation of strains that may be of crucial importance to production. Again, collections may provide a culture preservation service in which strains are expertly preserved and returned to the customer for storage. This may be useful both to industry, which is assured of a reliable supply of a "standard inoculum" for production purpose, or to research in which certain experiments may require the use of aliquots of a single population of cells.

With the present rapid developments in biotechnology there is an increasing need to patent processes or, more recently, genetically engineered strains, and there is a need for recognized depositories. Many culture collections have now become accepted as International Depository Authorities and are able to provide the legal requirements for the purpose. Acceptance of collections for this role demands certain assurances regarding permanence, impartiality and the existence of appropriate scientific expertise and facilities.

It is clear from a consideration of these varied activites that culture collections are not merely collections of cultures but provide a number of essential services for the general support of microbiology. In addition, it has always been considered important for collections to carry out research programmes that complement their interests. This not only ensures awareness of developments in their own fields but takes the maximum scientific advantage of having large numbers of authenticated microorganisms available for study.

Clearly, one area of research of particular interest to culture collection laboratories is taxonomy and its relevance to identification procedures. Again, preservation methods (particularly cryopreservation) and problems related to cell survival and strain stability are of fundamental importance. There is, moreover, a growing awareness of the extent to which many of the microorganisms maintained in the culture collections have not been fully characterized and collections have a developing interest in screening procedures to enable the research and industrial potential of the cultures maintained to be fully explored. The diversity of the types of microoganisms maintained, the interest of parent institutes and the changing requirements of microbiology lead to substantial variety in the kinds of research undertaken by collections.

I. THE UK NATIONAL SERVICE CULTURE COLLECTIONS

The United Kingdom maintains a decentralized network of national collections of microorganisms housed in research institutes with related interests. There are ten individual culture collections (see Table I) maintaining algae and protozoa, bacteria (medical, veterinary, industrial, marine, plant pathogenic), fungi (medical, industrial, wood-rotting), yeasts, dairy microorganisms and animal cells. The addresses of the collections are listed in Appendix II. These collections vary substantially in size and in types of ancillary activities (Table II); they frequently reflect the interests of the parent institutes and have developed in different ways to meet different requirements.

Table I: UK national culture collections.

Culture Collection	Acronym	Parent Institute	Type of Culture	Holding
NC of Type Cultures	NCTC	Central Public Health Laboratory, London	Bacteria (medical, veterinary)	4000
NCs of Industrial and Marine Bacteria	NCIMB	Torry Research Station, Aberdeen	Bacteria (general, industrial, marine)	5000
Collection of the Commonwealth Mycological Institute	CMI	Commonwealth Mycological Institute, Kew, Surrey	Fungi	10000
Culture Centre for Algae and Protozoa	CCAP	Institute of Terrestrial Ecology, Cambridge	Algae, protozoa	2000
NC of Yeast Cultures	NCYC	Food Research Institute, Norwich	Yeasts, other than known pathogens	2000
NC of Dairy Organisms	NCDO	Institute for Research in Dairying, Reading	Bacteria (general, dairying)	2000
NC of Pathogenic Fungi	NCPF	Mycological Reference Laboratory, London	Fungi pathogenic to humans and animals	830
NC of Plant Pathogenic Bacteria	NCPPB	MAFF Laboratory, Harpenden	Bacteria (plant pathogens) and associated phage	3000
NC of Wood Rotting Fungi	NCWRF	Princes Risborough Laboratory, Building Research Establishment	Wood-rotting basidiomycetes	620
National Animal Cell Culture Collection	NACCC	PHLS Centre for Applied Microbiology and Research, Porton Down	Animal cells	200
				29,650

Adapted from the UKFCC/UKNC Report on the National Culture Collections (Kirsop and Bousfield, 1982).
NC = National Collection.

Table II: Services provided by the UK service culture collections.

Collection	Advisory	Identification	Freeze-drying	Safe-deposit	Patents
NCTC	+	+	+	–	+B
NCIMB	+	+	+	–	+B
CMI	+	+	+	–	+B
CCAP	+	+	–	–	+B
NCYC	+	+	+	+	+B
NCDO	+	–	–	–	+*
NCPF	+	+	–	–	–
NCPPB	+	R	–	–	–
NCWRF	+	–	–	–	–

B—International Depository Authority (IDA) under Budapest Treaty.
*—Application to become IDA in progress.
R—Restricted service.

Adapted from the UKFCC/UKNC Report on the National Culture Collections (Kirsop and Bousfield, 1982).

A deliberate decision not to centralize the collections was taken in 1947 at a Commonwealth Specialist Conference on culture collections. It was felt that the collections would benefit scientifically from the close association with the activities of the parent institutes and a closer liaison would be established between collections and users. In order to provide the necessary co-ordination between the collections a UK National Committee for Culture Collections (UKNC) was set up. This committee meets once a year to discuss topics of common interest to the individual collections.

In 1975, a UK Federation for Culture Collections (UKFCC) (see Appendix II) was formed to promote and aid culture collections and to encourage communication between collections and their users. Membership is open to anyone with an interest in culture collections and their activities. Since its formation, the Federation has arranged one or two scientific meetings each year on such topics as cryopreservation, the stability of industrial cultures, quality control and the role of culture collections. It has run courses on preservation methods and has been responsible for a number of publications. The Federation is concerned with ensuring the continuation of existing collections and with encouraging collaboration between collections, their administrators and users.

In view of the benefits to be derived from international co-operation the International Association of Microbiological Societies

(IAMS) approved the formation in 1966 of a "Section on Culture Collections". In 1970 this Section was reorganized as the World Federation for Culture Collections (WFCC) (Appendix II). The WFCC has since produced a "World Directory of Collections of Cultures of Microorganisms", the second edition of which is now available (McGowan and Skerman, 1982). The Federation holds scientific meetings, runs training courses, and acts as an international forum for matters relating to culture collections.

3

General Introduction to Maintenance Methods

J. J. S. SNELL
Division of Microbiological Reagents and Quality Control
Central Public Health Laboratory
London, UK

I. CHOICE OF MAINTENANCE METHOD

A wide variety of techniques are available for the preservation of microorganisms and it may be difficult to choose the method most suitable for a particular need. This chapter provides a summary of the main features of various available methods. The reader is also directed to the reviews of maintenance methods by Lapage and Redway (1974) and Lapage *et al.* (1978). All methods have their unique advantages and disadvantages and the choice of method for a particular use should be determined by relating the features of each method to the needs of the user. Features to be considered are described below.

MAINTENANCE OF MICROORGANISMS
ISBN 0 12 410350 2

A. Maintenance of Viability

Cell death may occur during the preservation process and there may be further losses during storage, eventually resulting in unacceptably low levels of viability. To avoid loss of the culture it must be resubjected to the preservation process. The method used should minimize loss of viability during processing and storage so that once preserved cultures will survive for long periods.

B. Population Change through Selection

A proportion of the cells in a population may die during the initial preservation process and this may appear of little significance to the user if high initial cell concentrations are used. However, the reduction in the number of viable cells may result in the selection of a resistant population of surviving cells and introduce the possibility of change in the characteristics of the preserved culture. The preservation method should therefore retain the greatest number of viable cells so that the surviving population resembles the original as closely as possible.

C. Genetic Change

Microorganisms are usually preserved because certain strain characters are of scientific or industrial significance. It is important that preserved strains do not lose important characters or gain others. Changes may occur during preservation through mutation or loss of plasmids so the preservation method should minimize the occurrence of these events.

D. Purity

Cultures preserved for most applications should remain pure and the preservation method should minimize the chance of contamination.

E. Expense

The cost of maintaining cultures includes the costs of staffing, equipment, materials and general facilities such as storage space and power supplies. The high capital cost of equipment for some methods such as freeze-drying may be offset against reduced staffing costs, since the long-term stability of cultures reduces the need for frequent manual intervention.

F. Number of Cultures

The main factor to be considered in relating the number of cultures to the choice of method is the amount of operator time required for initial preservation and subsequent manipulation. A method found suitable for preservation of a small collection may be too labour intensive when the number of strains increases. Choice of method for larger numbers of cultures may be affected by the amount of storage space available.

G. Value of Cultures

The consequences of loss of a culture should be considered in choosing a preservation method. Important cultures should be preserved by methods that minimize the risk of loss and for complete security more than one method should be used. For less important cultures criteria such as cost may carry more weight.

H. Supply and Transportation of Cultures

If cultures are to be distributed, replicates of the culture are needed. These may be prepared as required or prepared in bulk and stored for later distribution. The convenience of either approach depends on the preservation method used and the number of cultures to be distributed. If cultures are to be supplied through the post they must be in a form suitable for packing and must survive the delays and conditions likely to be encountered. Strict adherence to national and international postal regulations should be made. Copies of these regulations may be obtained from national postal authorities.

I. Frequency of Use of Cultures

Some cultures such as assay strains, industrial production strains or those used for quality control may be used frequently within a laboratory. In these cases, ease of resuscitation and the risk of contamination of stock cultures need to be considered. Other cultures may be used rarely and other factors may be of more importance.

No single method of preservation fulfils all these criteria and selection of a method will be a process of compromise. It is important to recognize that microorganisms differ in their tolerance to various preservation methods and, unless a collection is very specialized, it is unlikely that a single method will provide optimum conditions for all strains. Again, the choice will depend on a balance of advantages and disadvantages.

Most of the methods described in this chapter are discussed in greater detail in other sections of the manual and their description here is limited to an outline of the general principles involved and an indication of how the methods relate to the criteria discussed above. Some methods have been applied only to restricted groups of microorganisms and it is not possible to comment on their general usefulness. However, it may well be that methods developed for a particular group may have wider applications and it is hoped that this manual will encourage investigations into the suitability of methods for microorganisms other than those for which they were originally described.

II. METHODS

A. *Subculture*

This method consists of inoculation of a suitable medium contained in a tube or bottle, incubation at an appropriate temperature to obtain growth, and storage under suitable conditions. The process is repeated at intervals that ensure the preparation of a fresh culture before the old one dies. The time that may be allowed to elapse between subcultures without risk of losing the culture depends primarily on the particular microorganism. Thus many bacteria such as staphylococci and coliforms will survive for several years, whereas the more delicate *Neisseria* spp. may require subculture after only a few weeks.

This method is inexpensive in terms of equipment but may be labour intensive if organisms that require frequent subculture are maintained. Cultures are easily resuscitated since a further subculture is all that is required to obtain an active culture. The method is applicable to a wide range of microorganisms. However, contamination is a major problem and this risk occurs with each subculture. Apart from the undesirability of mixed cultures, any contaminants may outgrow and kill the original culture. Contamination may be reduced by sound bacteriological technique and by pre-incubation of media before use. A two-tube method, where one is kept as a seed stock and another as a working culture, reduces the risk of contamination of the stock through frequent manipulation.

The risk of mislabelling or transposition of cultures is high, particularly where strains need frequent subculture. The risk increases with operator fatigue and can be alleviated to some extent by placing the numbered containers in random order thus maintaining the concentration of the operator. Labels should be printed or typed, as handwriting is open to misinterpretation and numbers may become completely altered over time.

Loss of viability is a constant hazard with this method. If strains with different survival characteristics are maintained an established protocol is essential to ensure timely subculture of all strains. A system for this purpose has been described by Skerman (1973). Loss of cultures is usually sporadic but whole batches may be lost because of such factors as faulty media. Dehydration may occur through imperfections in the seal of the container. Screw caps are not entirely free from the problem and plastic caps in particular may fail to seal a high proportion of the containers.

Cultures preserved by this method are very prone to loss of stability of characters and the risk of change increases with the frequency of subculture. A large inoculum reduces the risk of selection, but increases the risk of contamination. Neither is the method particularly convenient for the distribution of cultures, since a subculture must be prepared and checked for purity before despatch and postal conditions may adversely affect viability. The media used may affect survival time. Although a great variety of media have been used for the storage of various microorganisms there is a tendency to use unenriched media with limited nutrients. Storage in plain water has been found suitable for some organisms (Berger, 1970). For most microorganisms excess carbohydrates should not be included in the medium since the acid produced might kill the culture.

Storage periods can be extended by reducing the metabolic rate of the microorganisms. This may be achieved by restricting the availability of air to the culture; liquid paraffin, as a layer on the surface of the culture, has been used extensively for this purpose, especially with fungi (Onions, 1971). Metabolic activity may also be reduced by lowering the temperature and storage at 5°C has been widely used for a variety of microorganisms. However, not all microorganisms survive longer in culture at lower temperatures; *Neisseria* spp., for example, appear to survive better at 37°C.

B. Drying

Desiccation as a means of preserving microorganisms has been extensively used. A wide variety of methods exist but all consist essentially of removal of water and prevention of rehydration. The methods have been most widely applied to fungi, which appear more resistant to drying than other groups of microorganisms. Some yeasts and selected bacteria have also been successfully preserved by drying.

1. Sand, soil, kieselguhr, and silica gel

Sporulating fungi survive drying in soil (Fennell, 1960) and various species have been stored for periods of up to 5 years without change in characters (Atkinson, 1954). Selected yeasts, fungi and bacteria have been successfully dried on silica gel (Kirsop, this volume, ch. 12; Grivell and Jackson, 1969; Onions, 1971). Survival of bacteria for several years without change in characters has been reported by Grivell and Jackson (1969), but changes have been detected in some strains of yeast (Kirsop, this volume, ch. 12).

2. Paper strips or discs

Storage on paper has been successfully used with some yeasts (see Kirsop, this volume, ch. 12, IVA) and staphylococci (Coe and Clark, 1966). After drying the strips or discs may be stored in foil packets in airtight containers or between strips of self-adhesive plastic (Coe and Clark, 1966), providing an inexpensive method for the postage of large numbers of cultures.

3. Predried plugs

Various materials such as starch, peptone or dextran have been used to make predried plugs on to which small volumes of suspension of organisms are dropped before drying and storing under vacuum. This method has been successfully applied to delicate bacteria (*Neisseria gonorrhoeae* and *Vibrio cholerae*) which are difficult to freeze-dry (Annear, 1956).

4. Gelatin discs

In this method, originally described by Stamp (1947), organisms are suspended in a nutrient gelatin medium and drops are allowed to solidify in petri dishes. The drops may be dried or freeze-dried and the resulting dried discs stored over silica gel or phosphorous pentoxide. Various additives have been used to supplement the nutrient gelatin. A variety of bacteria have been successfully preserved by this method with survival over several years. (Snell, this volume, ch. 6; Obara *et al.*, 1981; Yamai *et al.*, 1979).

None of the above methods appear to be universally applicable and in most cases they have been used for particular groups of microorganisms or for rather specialist applications. It is therefore difficult to evaluate these methods fully in terms of the criteria

.discussed earlier as in many cases the relevant information is not available. Long-term viability appears moderately good and can certainly be measured in years. Stability of strain characters has not always been examined and appears to be very strain-specific. Contamination is certainly likely to be less of a problem than with serial subculture. Capital equipment costs are small and none of the methods appears unduly labour intensive in view of the expected survival times. Because of their technical simplicity, many of the methods would appear suitable for storing large numbers of cultures. Information on the reliability of these methods is restricted to a limited range of microorganisms and users would be well advised to gain first-hand experience before applying them to important cultures. Distribution would appear to be no problem as most of the methods lend themselves to batch production. Most of the methods appear specially suitable for storage of frequently used cultures since aliquots of dried material can be removed from containers without great risk of contamination.

C. Freeze-drying

Freeze-drying is a process in which water is removed by evaporation from the frozen sample. Organisms are suspended in a suitable medium, frozen and exposed to a vacuum. The water vapour removed is trapped either in a refrigerated condenser or phosphorous pentoxide. After drying the microorganisms are stored under vacuum or in an inert gas, most commonly in individual vials or ampoules. Two types of commercial freeze-dryer are in common use—the centrifugal and the shelf.

In the centrifugal dryer, the suspension is frozen by loss of latent heat associated with evaporation by vacuum. To increase the surface area and to avoid frothing of the suspension due to removal of dissolved gases before freezing is complete, the suspension is centrifuged during the initial stages of drying. Glass ampoules, which may be plugged, are used for centrifugal drying. After primary drying has been completed the ampoule is constricted and placed upon a manifold for secondary drying. Vacuum is applied and further water is removed before flame-sealing the ampoule, either under vacuum or inert gas.

In the shelf dryer the suspension is prefrozen before vacuum is applied. Freezing may be carried out by cooling the shelves in the freeze-dryer or by prefreezing in a deep-freeze. Glass vials are used for shelf drying and may be stoppered automatically in the machine.

It is not necessary to use a manifold for secondary drying as the complete drying programme is carried out continuously without further manipulation.

In general, centrifugal drying has achieved greater popularity than shelf drying. One advantage of centrifugal drying is that a cotton-wool plug can be inserted in the ampoule eliminating cross-contamination and acting as a filter to prevent scatter of micro-organisms into the environment when the ampoules are opened. A further advantage is that the glass seal of ampoules is both air and moisture-tight for long periods, whereas vials are sealed with various formulations of rubber which may allow access of air and water vapour either through the natural permeability of the bungs or through leakage of the seals.

Freeze-drying has been widely used to preserve many different microorganisms (Lapage and Redway, 1974) and is widely applicable to yeasts, fungi, bacteria and some viruses. It is less applicable to algae and is unsuitable for protozoa.

A variation on freeze-drying termed "L-drying" has been described (Annear, 1958; Lapage *et al.*, 1970; Bousfield, this volume, ch. 9). In this method, as in freeze-drying, suspensions of organisms are dried in ampoules under vacuum, but the vacuum is adjusted to allow rapid drying without freezing. The method has been used successfully with some species of bacteria that do not survive freeze-drying (Bousfield, this volume, ch. 9).

Although simple freeze-drying protocols can be established by trial and error, the process allows great refinement of control over a number of parameters and it is possible, if time permits, to optimize the process for the particular microorganisms to be preserved. The physical factors involved in the freeze-drying process have been discussed by Meryman (1966). The growth phase of the culture, temperature of growth, composition of suspending medium, rate of freeze-drying, final temperature of freezing, rate and duration of drying, and final moisture content are all factors which can be controlled and adjusted.

The advantages of freeze-drying are: suitability for batch production and distribution; maintenance of viability during storage (50 years or more for some microorganisms) without need for further attention; and undemanding storage requirements. For these reasons freeze-drying has found popularity in the service culture collections. The method suffers from some disadvantages. The capital cost is high if commercial equipment is purchased. Although a choice of machines is available commercially, equipment does not have to be sophisticated or unduly expensive since a vacuum pump, a manifold and a moisture trap are the only essential requirements. The freeze-drying process is fairly labour intensive but as large batches can be prepared the labour hours per ampoule may be quite low.

Although, in general, stability of characters is good some selection appears to take place during freeze-drying and up to a thousand-fold drop in viability is not uncommon with more susceptible microorganisms. Selection may be amplified if serial batches are prepared so that batch 2 is prepared from batch 1, batch 3 from batch 2, and so on. This effect can be avoided by reserving sufficient ampoules from batch 1 to act as a seed stock for subsequent batches. In addition to population changes, genetic change and loss of plasmids may also be experienced during freeze-drying of some species. From the user's point of view freeze-dried cultures are time-consuming to open and resuscitate and several subcultures may be needed before organisms regain their usual morphological and physiological characteristics.

D. Freezing

In preservation by freezing, water is made unavailable to the microorganisms by freezing it and the dehydrated cells are stored at low temperatures. Damage may be caused to the cells both during the cooling stage and the subsequent thawing. This may be caused either by the concentration of electrolytes through the removal of water as ice or by the formation of ice crystals which may damage cellular integrity. Attempts to limit this damage may be made by adjustment of cooling and warming rates and the addition of various cryoprotectants such as dimethyl sulfoxide or glycerol to the cell suspension.

Methods can be broadly classed according to the storage temperature used. Temperatures of $-20°C$, $-30°C$, $-40°C$, $-70°C$, $-140°C$ and $-196°C$, have all been used but, in general, temperatures above $-30°C$ give poor results, due to the formation of eutectic mixtures exposing cells to high salt concentrations. Storage at $-70°C$ has been used for a variety of different microorganisms including bacteria, fungi, mycoplasma, protozoa and viruses. Storage at $-140°C$ (nitrogen vapour phase) and $-196°C$ (nitrogen liquid phase) are being used increasingly and successful results have been achieved for microorganisms that cannot be preserved in other ways.

1. Storage on glass beads at $-70°C$

A novel approach to storage at low temperatures was suggested by

Feltham *et al.* (1978; see also Jones *et al.*, this volume, ch. 5). In this method storage at −70°C on small glass beads in a glycerol-suspending medium has been successfully used for a wide variety of bacteria. The method is very quick and easy and requires no subsequent manipulation during storage. Each glass bead provides material for one subculture and allows large batches to be stored in minimal space. The method is ideally suited to storage of large in-house collections.

A disadvantage of the method is the high capital cost of a −70°C freezer. In addition, provision must be made to safeguard against mechanical failure or prolonged interruption of electrical supply. Devices are available which provide automatic flushing with cold nitrogen vapour when a rise in temperature of the freezer occurs. Alternatively, a duplicate collection may be kept. The method is not well suited to frequent distribution of cultures as subcultures need to be made before issue.

2. Storage in liquid nitrogen
Storage in the liquid or vapour phase of nitrogen is the most universally applicable of all preservation methods. Fungi, bacteriophage, viruses, algae, protozoa, bacteria, yeasts, mammalian cells and tissue cultures have all been successfully preserved. Although with some microorganisms a high proportion of cells in a population may die on cooling and warming and population changes may occur, virtually no further loss occurs on storage. Losses may be reduced by the use of cryoprotectants and adjustment of growth conditions and the rate of cooling and warming. For a detailed discussion of these factors see Morris (1981, and this volume, ch. 13) and James (this volume, ch. 14). Although characters generally appear to be preserved without alteration, there have been some reports of nuclear and plasmid change (Calcott and Gargett, 1981; Williams and Calcott, 1982). Current knowledge suggests longevity and stability of cultures with this method is higher for most microorganisms than following freeze-drying, and storage in liquid nitrogen is nowadays the method of choice for preservation of valuable seed stock cultures. Storage in liquid nitrogen may be the only suitable method for long-term preservation of microorganisms that will not survive freeze-drying.

The method has some disadvantages however. Liquid nitrogen evaporates and must be replenished regularly. Failure to do this, either through laboratory mishap or interruption of deliveries of liquid nitrogen through industrial action, may cause the loss of an entire collection. The capital cost of equipment is high, but the process is not labour intensive. There is some risk of explosion if glass containers are kept in the liquid phase, as liquid nitrogen may penetrate through imperfect seals and expand rapidly when the container is warmed. Storage in the vapour phase removes this hazard, but the higher storage temperature may be considered less satisfactory for valuable strains. The method is not very convenient for the distribution of cultures as subcultures have to be prepared. Storage space may become a problem particularly if working cultures are stored in addition to seed stocks, but various methods have been developed to reduce the storage space required (straws, Kirsop, this volume, ch. 12; glass beads, Bousfield, this volume, ch. 9; glass capillaries, Hippe, this volume, ch. 10, and James, this volume, ch. 14).

4

Maintenance of Bacteria by Freeze-drying

R. H. RUDGE
National Collection of Type Cultures
Central Public Health Laboratory
London, UK

I. INTRODUCTION

Freeze-drying, or lyophilization, is a process in which water vapour is removed directly from a frozen product by sublimation. It has been used for many years to preserve a wide variety of biological materials and among the numerous publications on the subject are reviews by Fry (1954, 1966), Heckly (1961, 1978), Hill (1981), Lapage and Redway (1974), Lapage *et al.* (1970) and Muggleton (1963), which cover freeze-drying of bacteria.

MAINTENANCE OF MICROORGANISMS
ISBN 0 12 410350 2

Unlike drying from the liquid phase, freeze-drying causes little shrinkage and results in a completely soluble product that is easily rehydrated. Chemical changes are minimized by preventing concentration of solutes and also by virtue of the lowered temperature, which reduces the rate of chemical reaction. One major advantage of freeze-drying over many other preservation methods is that material can be kept stable over a period of many years without the need for special storage conditions. Distribution of cultures is also simplified as no further preparation is necessary before despatch. There are many different methods of freeze-drying, but the techniques covered here are limited to those used by the National Collection of Type Cultures (NCTC).

II. DETAILS OF METHOD

1. Pre-drying Culture Preparation

It is desirable that the best growth possible is obtained prior to drying and that it is in an easily harvested form. Thus cultures are grown in 150 × 19 mm tubes of suitable sloped medium, commonly nutrient or blood agar, using one tube per ten ampoules to be dried. If necessary, liquid cultures can be used, though these need to be centrifuged in order to concentrate the cells before preparation of the drying suspension. The optimal stage of growth for harvesting may vary but, in general, late logarithmic phase cultures prove suitable.

2. Suspending Fluids

(i)   Inositol serum (Redway and Lapage, 1974)
       meso-Inositol (Koch-Light)   5 gm
       Horse serum No. 3 (Wellcome)   100 ml
Sterilize by filtration and aseptically distribute 5 ml volumes into sterile bijoux bottles.

(ii)   Inositol broth
       Nutrient broth powder No. 2 (Oxoid)   2.5 gm
       meso-Inositol (Koch-Light)   5 gm
       Distilled water   100 ml
Distribute 5 ml volumes into bijoux bottles and sterilize by autoclaving at 121°C for 15 min.

C. Preparation of Suspension

A suitable suspending fluid is essential to prevent overdrying and to protect the bacteria from mechanical and chemical damage both during drying and storage. The NCTC uses 5% inositol serum routinely for all organisms except enterobacteria, for which 5% inositol broth is used in order to avoid possible immunological change or damage.

Harvesting is carried out by adding 1–2 ml of suspending fluid to each slope of culture and gently rubbing off the growth with a Pasteur pipette before emulsifying into a uniform suspension. This should be done carefully to avoid creating aerosols. With larger batches, the growth from several tubes is pooled prior to filling ampoules.

4. Preparation of Ampoules

Neutral glass freeze-drying ampoules, 100×7–7.5 mm (Edwards High Vacuum), are acid washed by soaking in 2% hydrochloric acid overnight, then thoroughly rinsed in tap water and finally distilled water. For each culture to be dried, labels are prepared by stamping or typing the required identification on blotting paper; a 5×30 mm strip is placed in each ampoule. Ford's gold label 428 mill paper, 140 gm m^{-2} (Universal Stationers), is used in the NCTC and different colours are available for batch coding. Ampoules are then plugged with cotton wool and sterilized by autoclaving at 126°C for 20 min.

5. Filling Ampoules

Ampoules are filled with the bacterial suspension using a Pasteur pipette; approximately 0.1–0.2 ml is delivered to the bottom of each ampoule, taking care not to contaminate the sides or top. As the ampoules are filled, the tops are flamed thoroughly and the cotton wool plugs replaced, though these are later discarded when all the ampoules are ready for loading on to the dryer.

6. Primary Drying

This stage involves removal of all the water in the preparation that can be frozen. The suspensions are initially frozen by evaporative freezing under reduced pressure while centrifuging to prevent frothing due to evolution of dissolved gases. Because of the high rate of vapour flow during this process, obstructions such as cotton wool plugs should be avoided; instead, caps of gauze or cotton placed over each ampoule or group of ampoules, while not providing an absolute bacterial filter, will limit any contamination and contain any loose flakes of freeze-dried material. Primary drying is continued for a minimum of 3 h.

7. Constricting Ampoules

On completion of primary drying, the ampoules are constricted to facilitate later sealing. It is advisable that this stage is carried out as quickly as possible to avoid prolonged exposure of the ampoules to air. Sterile plugs of non-absorbent cotton wool, previously auto-claved *in situ* in empty ampoules or tubes, are inserted approximately 15 mm into each ampoule and the top section, which has been handled, is cut off. The plug is pushed with a ramrod halfway down the ampoule and then, using a narrow flame and taking care not to char the cotton wool, the ampoule is constricted above the plug to produce a short capillary section of about 2 mm diameter. In the NCTC, this is done with a "Flair" handtorch (Jencons Scientific), using natural gas and oxygen to produce a very hot, slim flame. Alternatively, a fishtail burner or semi-automatic ampoule con-strictor (Edwards High Vacuum) can be used.

8. Secondary Drying

During this stage, the water that remains unfrozen, bound to the material by adsorption, is removed to leave a residual moisture content of around 1%. Compared with primary drying, desorption is a slow process and is generally carried out using a chemical desiccant such as phosphorus pentoxide (P_2O_5) to trap the small amounts of water involved. Refrigerated vapour traps are less efficient, but can be used provided a condenser temperature of below $-50°C$ is maintained.

9. Sealing

When secondary drying is complete, ampoules are sealed *in situ* on the dryer manifold while still under vacuum. The "Flair" handtorch (Jencons Scientific) is well suited to this, but alternatively a crossfire burner (Edwards High Vacuum) or small domestic blowtorch, e.g. Ronson, may be used. If the constriction is too wide, or the flame held too close to the body of the ampoule, a hole may be produced in the ampoule as the melted glass is sucked inwards. Ampoules damaged in this way must be discarded.

10. Hazardous Pathogens

Category B and other hazardous pathogens are dried in the NCTC by a slightly different method, which avoids evaporative freezing and centrifugation, and employs cotton-wool plugs throughout the whole drying process. All operations are carried out in a self-contained Category B room, and ampoules are filled and subsequently opened in an exhaust protective cabinet.

After filling, ampoules are re-plugged with loose, non-absorbent cotton-wool plugs, which are then trimmed and pushed halfway down the ampoule. Suspensions are quick-frozen by covering ampoules while they are sloped with crushed solid CO_2 and the frozen ampoules are transferred quickly to the freeze-dryer and subjected to a vacuum before they can thaw. Drying then proceeds as above, with the exception that the ampoules do not need to be plugged prior to constriction.

11. Vacuum Testing of Ampoules

For locating leaks in ampoules on the secondary drying manifold, and also for checking maintenance of vacuum in ampoules after storage, a useful item of equipment is a high frequency spark tester (Edwards High Vacuum). This demonstrates a satisfactory vacuum by producing a pale-blue/violet glow within the ampoule, whereas a poor vacuum is indicated by a deep-purple glow or else no discharge at all. The tester should be used with care, however, avoiding the bottom and sealed tip of the ampoule as these more fragile areas can be punctured by the spark.

12. Opening Ampoules

Ampoules can be opened safely by making a score mark on the glass with a file or diamond at a point near the middle of the cotton-wool plug and then applying a red-hot glass rod or pipette tip to this mark. This should produce a crack encircling the ampoule; if only a short crack is produced, tap gently at this point to complete the encirclement. After allowing air to seep in, filtered by the plug, the tip of the ampoule can be removed; this and the plug are then discarded, treating both as infected material. The open end of the ampoule is flamed and a sterile cotton-wool plug inserted.

13. Reconstitution

With a Pasteur pipette, approximately 0.5 ml nutrient broth is added to the ampoule and the contents mixed carefully to avoid frothing. The suspension is then used to inoculate suitable media including, if possible, an agar plate in order to detect any contaminants introduced during opening.

14. Viability Counts

Each batch of cultures dried in the NCTC is checked routinely by counting the suspension before drying and again immediately after drying. This provides not only a viability check and a measure of the loss during drying, but also serves as a check for purity of the dried culture. Unless the count is low enough to warrant more frequent checks, further counts are made after 1 year, 5 years and thereafter at 5-year intervals. Cultures are re-dried when the count falls below an acceptable limit.

Counts are performed using a modified Miles and Misra (1958) technique. The culture is rehydrated in 1 ml nutrient broth to provide a nominal 10^{-1} dilution, and from this a series of tenfold dilutions is prepared up to 10^{-6}. Using a pipette calibrated to deliver drops of 0.02 ml (Laboratory Environmental Supply Associates), three drops of each dilution are delivered onto an agar plate of suitable medium. Results are generally expressed in terms of the highest dilution to yield growth, as this is adequate for the purpose of comparison and there is no need to calculate the actual count per ampoule.

III. OPERATION OF MACHINES

A. *Operation of Edwards EFO3 Freeze-dryer (Basic Unit)*

(1) Ensure that all valves are closed and switches off.
(2) Load the desiccant trays with P_2O_5, using a minimum of 5 gm for each 1 ml volume to be dried.
(3) Place trays in vapour trap and secure cover.
(4) Load centrifuge head and remove ampoule plugs.
(5) Cover each ampoule or group of ampoules with gauze or cotton caps.
(6) Place centrifuge head on chamber spindle and position chamber jar on baseplate.
(7) Switch on centrifuge.
(8) Switch on Pirani gauge and select head 1.
(9) Switch on vacuum pump.
(10) Slowly turn chamber isolation valve until fully open.

(11) Centrifuge for 5–10 min or until vacuum reaches approximately 0.2 torr (0.26 mbar). Ampoules should now be frozen and the centrifuge can be switched off.
(12) Continue drying for a minimum of 3 h, after which time primary drying should be complete.
(13) Close chamber isolation valve and slowly open chamber air-admittance valve.
(14) When the vacuum has been released remove chamber jar and unload centrifuge head.
(15) Plug and constrict ampoules.
(16) Fit secondary drying manifold to coupling on chamber top and position chamber on baseplate.
(17) Place ampoules on manifold, twisting the ampoules slightly to ensure a good seal.
(18) Close chamber air-admittance valve and slowly open chamber isolation valve.
(19) Ensure that Pirani gauge reaches a pressure <0.04 torr (0.05 mbar); if not, check ampoules for leaks using a high-frequency spark tester and replace faulty ampoules with blanks.
(20) Continue secondary drying overnight, then flame-seal ampoules under vacuum.
(21) Switch off dryer and open air-admittance valves.

The older Edwards model 5PS freeze-dryer also follows a similar sequence of operations.

B. Operation of EFO3 Freeze-dryer (Refrigerated Version)

1. Primary drying

Operation is identical to that of the basic unit except for stages:

(2) Open condenser drain valve and allow any condensate to empty. Close valve.
(3) Switch on refrigerator and allow condenser coils to reach at least −40°C.

2. Secondary drying

If continuing to use the condenser coils to complete drying, the sequence of operations remains as in IIIA. However, if using a chemical desiccant for secondary drying, the following steps should be included between stages 14 and 15:

(a) Switch off refrigerator.
(b) Switch off vacuum pump, open trap air-admittance valve and remove trap cover.

(c) Defrost condenser by filling with warm water and leaving until ice has melted.

(d) Open drain valve and wipe condenser dry. It is important that this is dried thoroughly before using P_2O_5; if necessary, use a hot-air blow-dryer. Close drain valve.

(e) Load desiccant tray with P_2O_5 and place tray on top of condenser coils.

(f) Refit and secure trap cover. Close trap air-admittance valve.

(g) Switch on vacuum pump.

The model EFO3 freeze-dryer has since been replaced by the Edwards EF4 Modulyo. However, apart from the lack of a chamber isolation valve, which necessitates switching off the vacuum pump during operations between the primary and secondary drying stages, this newer model remains similar in use to the refrigerated EFO3.

IV. STORAGE CONDITIONS

Stability of freeze-dried cultures stored in the dark at normal room temperatures is generally very good, although there may be some decrease in shelf life if ambient temperatures are consistently high. This difference can be utilized in the accelerated storage test, in which the viability loss of cultures held at higher temperatures is used to predict the shelf life at normal storage temperature (Mitic *et al.*, 1974). For most cultures, storage at room temperature will prove adequate, though some of the more delicate organisms (see section VI) may benefit from storage in a refrigerator or freezer.

V. GENERAL NOTES

(1) Used P_2O_5 is best disposed of by first leaving to hydrate completely by exposure to air. The sludge that remains can then be dissolved by soaking in hot water.

(2) Pyrex ampoules should not be used for freeze-drying. Although stronger, they can be extremely difficult to open.

(3) When preparing ampoules, and again when filling, the tops should be examined for cracks or irregularities and any flawed ampoules discarded as these can splinter or leak when placed on the secondary drying manifold.

(4) The rubber manifold nipples can be *lightly* greased periodically, using Apiezon high vacuum grease (Edwards High Vacuum) to facilitate a good seal with the ampoules.

(5) Ampoule identification labels should include a 10 mm gap to the left of the number and this end is placed towards the bottom of the ampoule. This prevents the dried material obscuring the number and helps to ensure that the figures are always read from the same end and thus the right way up.

(6) A guide to completion of primary drying on the model EFO3 dryer can be provided by temporarily isolating the drying chamber from the vapour trap and vacuum pump; water vapour still evolving from the ampoules will be indicated on the chamber Pirani gauge as a rise in pressure.

VI. ORGANISMS SUCCESSFULLY PRESERVED

Table I lists bacterial genera that are routinely dried in the NCTC and is compiled from records of viability counts over the past 30 years. Figures given for the mean logarithmic count represent an average value of results obtained for a given number of strains in each genus and indicate the highest dilution to give single figure colony counts. Not all the 3500 NCTC strains are included in these figures, but only those from each genus which have been counted over the longest period. Although some batches that were discarded early because of low counts are therefore omitted, this bias is somewhat compensated for by the inclusion of older batches that were dried and stored under suboptimal conditions.

The majority of bacteria survive freeze-drying well, but a few species can sometimes give disappointing results. This may be due in some cases to difficulties in obtaining adequate pre-drying growth. Cultures which often prove more difficult than others include *Clostridium botulinum*, *Cl. chauvoei*, *Cl. novyi (oedematiens)*, *Cl. putrificum*, *Cl. skatologenes*, *Peptococcus heliotrinreducans* and *Spirillum serpens*. Additionally, some lesser problems may be encountered with *Bacteroides melaninogenicus*, *Haemophilus canis*, *H. suis*, *Leptotrichia buccalis*, *Mycobacterium microti*, and *Neisseria gonorrhoeae*.

Genus	No. of Species	No. of Strains	Mean logarithmic counts before drying (BD), after drying (AD) and after storage for various periods (in years)								
			BD	AD	1	5	10	15	20	25	30
Achromobacter	6	9	7.0	7.0	6.9	6.8	(5.5)	+			
Acinetobacter	2	10	7.0	7.0	6.9	6.8	6.7	(6.5)	+		
Actinobacillus	3	19	6.3	6.1	5.6	4.9	4.2	(2.9)			
Actinomadura	3	8	4.6	4.5	4.4	4.3	(3.8)	(3.0)			
Actinomyces	4	13	5.2	4.9	4.5	4.2	3.8	(3.0)	(3.0)	(3.0)	(3.0)
Aerococcus	1	14	5.8	5.7	5.6	5.6	5.5	5.5	(5.1)		
Aeromonas	4	9	7.0	6.7	6.7	6.2	(6.5)				
Alcaligenes	3	7	7.0	6.9	6.6	6.6	(6.3)	+			
Alteromonas	3	8	6.9	6.8	6.6	6.0	(5.5)				
Alysiella	1	1	5.0	5.0	5.0	5.0					
Anaerobic coccus	—	11	6.2	5.7	5.2	4.8	4.5	3.8	(3.5)		
Bacillus	27	70	5.8	5.5	5.4	5.2	5.0	4.8	4.3	(3.0)	+
Bacterionema	1	3	5.3	5.0	5.0	4.0	3.0				
Bacteroides*	5	12	6.4	6.0	5.2	4.4	(3.4)	(3.3)	(3.3)		
Beneckea	4	4	6.3	5.8	4.8						
Bifidobacterium	1	2	6.5	6.5	6.5						
Bordetella	3	26	7.0	6.9	6.7	6.6	6.4	(6.3)	(6.6)		
Brevibacterium	—	2	7.0	7.0	7.0	7.0					
Brucella	5	32	6.8	6.8	6.8	6.5	6.2	5.8	(5.2)		
Butyribacterium	1	1	6.0	6.0	6.0	6.0					
Campylobacter	3	7	6.3	5.4	5.4	5.1	(4.0)	(4.0)			
Capnocytophaga	2	2	5.0	4.0	3.5						
Cardiobacterium	1	3	7.0	6.3	5.7	(4.0)					
Cellulomonas	2	2	6.5	6.5	6.5	6.5	6.0	5.0	5.0		
Chromobacterium	3	19	6.7	6.1	5.5	5.2	4.8	4.2	(4.0)		
Citrobacter	2	11	7.0	6.9	6.8	6.6	6.5	6.0	(6.0)		
Clostridium*	18	73	5.0	4.6	4.5	4.2	3.9	3.8	(3.2)	+	+
Comamonas	1	3	7.0	6.7	6.7	6.3	5.7	(6.0)			
Corynebacterium	14	72	6.4	6.1	5.9	5.8	5.4	5.1	4.6	(4.1)	(3.2)
Cytophaga	1	1	7.0	6.0	6.0	6.0	6.0				
Dermatophilus	1	1	4.0	4.0	4.0	4.0	4.0				
Edwardsiella	1	3	6.7	6.3	6.0	6.0	6.0	(6.0)			
Eikenella	1	2	6.0	6.0	5.5	4.5					
Enterobacter	2	14	6.6	6.6	6.6	6.4	6.3	6.1	(5.6)	(5.0)	(5.0)
Erwinia	1	4	7.0	7.0	7.0	7.0	7.0				
Erysipelothrix	1	8	6.3	6.1	6.0	5.9	5.6	5.4	(5.5)		
Escherichia	3	36	6.9	6.5	6.3	5.9	5.8	5.6	5.3	(4.0)	(3.7)
Flavobacterium	3	13	7.0	7.0	6.9	6.9	(6.3)	(6.0)	(6.0)		
Francisella	1	1	6.0	6.0	6.0		5.0				
Fusobacterium	2	2	7.0	7.0	6.5	6.0					
Gardnerella	1	2	6.0	5.5	5.0	4.5					
Gemella	1	3	6.0	6.0	6.0	6.0	5.0				
Haemophilus*	7	19	6.3	5.8	5.1	4.6	3.8	+	+		
Hafnia	1	8	7.0	6.4	6.0	5.6	5.3	4.9	4.6	4.6	+
Kingella	3	8	5.3	5.3	5.0	5.0	(4.5)				
Klebsiella	6	57	6.8	6.8	6.7	6.6	6.5	6.1	(5.9)	(5.8)	
Kluyvera	2	2	7.0	7.0	7.0	7.0	7.0	7.0			
Kurthia	2	3	6.3	5.7	5.3	5.0	4.3				

Table I—continued

Genus	No. of Species	No. of Strains	BD	AD	1	5	10	15	20	25	30
			Mean logarithmic counts before drying (BD), *after drying* (AD) *and after storage for various periods (in years)*								
Lactobacillus	3	6	5.7	5.3	5.2	5.0	4.5	(4.6)	(4.6)		
Legionella	4	11	6.1	5.8	5.5						
Leptotrichia*	1	1	6.0	5.0	4.0	3.0	0				
Leuconostoc	1	1	6.0	6.0	6.0	6.0					
Levinea	1	2	7.0	7.0	7.0	7.0					
Listeria	4	15	6.3	6.3	6.1	5.9	5.8	(5.7)	(5.0)		
Micrococcus	3	23	6.1	6.0	6.0	6.0	6.0	6.0	(5.6)	(5.5)	(5.3)
Moraxella	12	25	6.0	6.0	5.8	5.4	(5.1)	(3.7)			
Morococcus	1	1	5.0	4.0							
Mycobacterium*	12	49	5.6	5.4	5.2	5.0	4.6	4.5	(4.0)	(3.7)	(4.0)
Mycococcus	2	2	6.5	6.5	6.5	6.5	6.5				
Neisseria*	9	41	6.6	6.1	5.6	4.9	4.1	(3.0)	(2.3)	+	+
Nocardia	5	10	6.1	6.0	5.9	5.9	5.7	+	+		
Pasteurella	5	28	6.9	6.8	6.2	5.8	5.6	5.4	(5.5)	+	
Pediococcus	1	2	6.0	6.0	6.0	6.0	6.0				
Peptococcus*	2	2	6.0	6.0	+	+					
Plesiomonas	1	3	7.0	6.0	6.0	5.7	5.7	(5.0)			
Propionibacterium	2	2	6.5	6.5	6.5	6.5	6.5				
Proteus	4	22	6.7	6.6	6.6	6.6	6.5	6.5	(6.3)		
Providencia	—	10	7.0	6.8	6.7	6.3	6.2	6.0	(5.6)	+	+
Pseudomonas	12	31	6.8	6.4	6.1	5.5	5.0	4.5	(4.1)	+	+
Ramibacterium	1	1	6.0	6.0	6.0	5.0	5.0				
Rhodococcus	5	14	6.5	6.5	6.4	6.3	6.1	6.0	(5.8)	(5.0)	(5.0)
Rothia	2	3	6.0	6.0	6.0	5.7					
Salmonella	—	131	7.0	6.5	6.2	5.8	5.4	5.1	4.9	4.7	4.4
Sarcina	1	1	7.0	7.0	7.0	7.0	7.0	7.0			
Serratia	1	12	6.9	6.8	6.8	6.8	6.5	6.4	(6.0)		
Shigella	4	63	6.9	6.7	6.5	6.1	5.8	5.6	5.5	5.1	+
Simonsiella	1	1	4.0	4.0	3.0	3.0	3.0				
Sphaerophorus	2	4	6.8	6.8	6.3	6.0					
Spirillum*	1	1	7.0	4.0	3.0	2.0					
Staphylococcus	3	48	6.3	6.1	6.1	6.0	5.9	5.8	5.6	5.3	(4.6)
Streptobacillus	1	2	4.5	4.5	4.0						
Streptococcus	12	75	6.0	5.8	5.7	5.5	5.1	5.0	4.7	4.5	(3.4)
Streptomyces	5	9	4.1	3.7	3.7	3.6	3.5	(3.8)	(3.5)		
Thermoactinomyces	1	1	4.0	4.0	4.0	4.0					
Vibrio	5	23	7.0	6.6	6.0	5.4	5.0	4.6	(3.8)	+	
Yersinia	2	20	6.6	6.6	6.4	6.4	6.2	5.9	+		
Zooglea	1	1	7.0	7.0	7.0	6.0	5.0				

Count figures in parenthesis are based on fewer than the indicated number of strains tested.
+ Indicates culture still viable, though an insufficient number of strains tested to provide a representative log count.
* Some species in these genera may prove difficult to freeze-dry and result in a relatively poor survival rate.

5

Maintenance of Bacteria on Glass Beads at −60°C to −76°C

D. JONES, P. A. PELL and P. H. A. SNEATH
Department of Microbiology
University of Leicester
Leicester, UK

I. INTRODUCTION

Storage of cultures in the range of −60°C to −80°C is possible in any laboratory because of the ready availability of commercial deep-freezers within this temperature range. The disadvantage of storing bacteria in this way is the damage caused by repeated freezing and thawing when subcultures are required. To overcome this problem a method based on the use of frozen bacterial suspensions with a cryoprotectant in glass beads was developed in our laboratory (Feltham *et al.*, 1978). The technique allows individual beads to be removed without thawing the whole sample. Over the 7 years of use in our and other laboratories, the method has proved to be a safe, reliable and simple procedure for the storage of a wide range of

MAINTENANCE OF MICROORGANISMS
ISBN 0 12 410350 2

bacteria (Feltham *et al.*, 1978; Pell and Sneath, 1983). The advantages of the method are:

(1) minimum preparation of materials is required;
(2) the method is simple to perform;
(3) many hundreds of strains can be stored for long periods in a small space;
(4) recovery of cultures is quick with little or no disturbance to other stored cultures;
(5) only the portion of culture removed is thawed; the bulk of the stock culture remains frozen;
(6) the beads thaw rapidly when placed on solid growth medium and recovery is immediate;
(7) with a suspending medium of suitable composition, the method can be used for most aerobic and anaerobic bacteria;
(8) stability of phenotype appears comparable to that achieved by freeze-drying;
(9) beads of different colours can be used to identify various categories of bacteria;
(10) viability and stability of cultures is not seriously impaired by breakdown of refrigeration for a few days.

II. DETAILS OF METHOD

The steps involved in preparing the bacterial suspensions are:

(1) sterilization of vials containing washed beads;
(2) growth of bacteria;
(3) suspension in an appropriate medium containing a cryo-protectant, such as glycerol;
(4) distribution of suspension to beads in vials;
(5) freezing of vials;
(6) checking of product.

1. Batch Size

The batch size prepared will depend on the intended use of the culture. About 20 to 30 prepared glass beads are placed in vials of 2 ml capacity. One culture may be distributed in more than one vial and it is advisable to prepare at least two vials per culture. One may then be used for routine recovery, the other as a reserve in case the first becomes contaminated or yields no growth. If two freezers are available one vial can go in each to guard against loss.

2. Preparation of Beads

Glass 2 mm embroidery beads (Creative Beadcraft Ltd) are washed in tap water with a detergent, followed by dilute HCl to neutralize alkalinity. The beads are then washed several times in tap water until the pH of the wash water is that of tap water. The beads are finally washed in distilled water, then dried at 45°C in an oven.

The suppliers produce beads of various colours which may be used to differentiate various groups of bacteria, according to the requirements of a particular laboratory; for example: animal pathogens, red; plant pathogens, green; special growth conditions required, blue; teaching strains, white.

3. Preparation of Sterilized Vials

About 20 to 30 prepared beads are placed in screw-cap glass vials of 2 ml capacity (R. W. Jennings Ltd). The vials are capped and sterilized by autoclaving at 121°C for 15 min. Quantities of autoclaved vials may be stored until required.

4. Preparation of Suspending Medium

For aerobic bacteria 10 ml quantities of 15% (v/v) glycerol in Nutrient Broth (Difco Laboratories) are prepared in Universal bottles and sterilized by autoclaving at 121°C for 15 min. For anaerobic bacteria BGP medium (Barnes, 1969) without agar but with 15% (v/v) glycerol is recommended. This medium is of the following composition:

Tryptone	(Oxoid)	10 g
NaCl		5 g
Beef extract	(Lab-Lemco, Oxoid)	3 g
Yeast extract	(Difco)	5 g
Cysteine hydrochloride		0.4 g
Glucose		1 g
Na_2HPO_4		4 g
Glycerol		150 ml
Distilled water		1000 ml

Dispensed in 10 ml quantities in Universal bottles.
Sterilized by autoclaving at 121°C for 15 min.

To freeze halophiles or alkalophiles the appropriate growth medium plus 15% (v/v) glycerol has proved satisfactory.

5. Growth of Bacteria

Bacteria should be grown on the most appropriate, non-selective, solid medium under the optimum growth conditions. The use of a solid, non-selective medium reduces the risk of contamination. The number of growth plates used for any culture depends on the batch size required and on the vigour of the organism. Experience has shown that loss of viability after repeated sampling of one vial is reduced markedly when thick bacterial suspensions ($>10^8$ organisms ml^{-1}) are used.

6. Labelling of Vials

Sterile vials are labelled for each organism. The appropriate culture collection number, the most suitable medium for recovery, date of preparation of material, and any other pertinent information is written on a self-adhesive label and stuck on the side of each vial. It is recommended that these labels be further secured by wrapping a layer of clear sticky tape completely around each vial. This method of labelling is less expensive than the special deep-freezer tapes. The black caps may be whitened with waterproof ink and the culture collection number written on the top with waterproof ink. This facilitates detection and retrieval of the appropriate vial when the organism is required.

7. Preparation of Bacterial Suspension

Strains incubated overnight under suitable conditions should be carefully inspected for contaminants. Approximately 1 ml of the appropriate sterile, suspending medium is aseptically pipetted onto the plate and using a wire loop the growth is emulsified with the broth to make a thick suspension.

8. Distribution of Suspension

With a sterile Pasteur pipette the bacterial suspension is aseptically dispensed into each of the two prepared vials. The suspension should be aspirated several times to ensure the air bubbles inside the beads are displaced by the bacterial suspension. After the beads are thoroughly wetted, the excess suspension should be removed from the bottom of the vial. Excess suspension left in the vial makes it more difficult to remove individual beads when required after storage.

An alternative method for mixing bacteria with the beads is to wet the beads with the appropriate suspending medium then agitate one loopful of the growth from the plate amongst the beads. Again, this should be done so that the air bubbles are displaced. This method is especially useful when only slight bacterial growth is available.

9. Freezing and Storage of Material

The vials are placed in trays of suitable size. We have used anodized aluminium sectioned trays (Denley Instruments Ltd). However, smaller trays could prove more convenient and a tray with a lid prevents accumulated freezer ice falling on the vials during manipulation.

The trays are placed in a commercial freezer (e.g. Model HLT 12V-75, Harris Mfg. Co. Inc.) capable of maintaining temperatures of −60°C to −80°C. A temperature of *circa* −70°C is recommended. Removal of single beads after storage is facilitated if the vials are frozen on their side or if the beads are tapped onto the side of the vial before freezing in the upright position.

If a number of bacterial cultures are stored by this method it is recommended that a record of the contents of each tray and the position of each tray in the freezer is recorded in a culture collection note book. This makes location and retrieval of cultures easier.

10. Recovery and Checking of Frozen Bacteria

As with any method of preservation the bacteria should be checked for viability, purity and retention of particular characteristics after freezing

The vial is removed from the freezer and one bead removed using a mini-spatula sterilized by flaming in alcohol and then cooling. The vial should be replaced immediately to prevent the remaining contents from thawing. The bead is rubbed over the surface of a suitable solid medium with a wire loop so that the bacterial inoculum is released. The plate is then incubated under appropriate conditions.

If a number of vials are removed at one time or if the freezer is located at some distance from the work bench, thawing may occur during transit. This problem can be minimized by immersing the vials in boxes containing solid CO_2 or by using a cold block of paraffin wax containing holes into which the vials fit. We favour the block of paraffin wax. This is made by pouring paraffin wax into a tin box; vertical holes of a size suitable to contain the vials are then drilled part way through the block. The block is kept in the freezing cabinet. Vials for sampling may be placed in the cavities and the block of wax carried to the laboratory bench; the cold wax keeps the vials frozen for up to an hour. Long-term storage of cultures in solid CO_2 is not recommended because absorption of CO_2 causes lowering of the pH of the cultures and this could be detrimental.

III. STORAGE CONDITIONS

The vials are stored in a commercial freezer at temperatures between $-60°C$ and $-76°C$. The actual temperature within this range does not appear to be very important. With any refrigeration unit there is always the risk of breakdown and emergency back-up facilities are expensive. Repairs are, however, usually conducted in a matter of hours or, at worst, a few days. Recent work (Pell and Sneath, 1983) indicates that bacteria frozen on beads can survive such breakdowns and those tested remained viable for a few days. It is recommended that if thawing has been in progress for more than a few hours, the collection should be re-preserved with newly grown bacteria.

IV. ORGANISMS SUCCESSFULLY PRESERVED

We have successfully preserved various species of *Actinobacillus*, *Haemophilus*, *Pasteurella*, *Yersinia*, *Vibrio*, a number of Entero-bacteriaceae, staphylococci, micrococci, lactobacilli, streptococci and coryneform bacteria. Other laboratories have successfully preserved alkalophiles, halophilic archaebacteria, propionibacteria and *Bacteroides* spp. The method has been in use for some 7 years but not all bacteria have been preserved for this length of time.

V. SHELF LIFE

In our experience to date, good levels of viability are maintained for up to 7 years. There is every reason to believe that the "shelf life" is, in fact, a good deal longer.

6

Maintenance of Bacteria in Gelatin Discs

J. J. S. SNELL
Division of Microbiological Reagents and Quality Control
Central Public Health Laboratory
London, UK

I. INTRODUCTION

Preservation of bacteria in the form of gelatin discs was first described by Stamp (1947). A harvest of bacterial growth is suspended in melted nutrient gelatin, drops of which are allowed to solidify in petri-dishes. The drops are dried, or freeze-dried, over a desiccant and the resultant flat discs are stored over silica gel. For use, a single disc is placed in warmed broth and the resulting suspension plated onto a suitable growth medium. The method is not particularly suitable for storage of numerous strains over long periods; however, it is invaluable for storage of a limited number of frequently used strains, such as those used for quality control of media or reagents. The essential advantages of the method are:

MAINTENANCE OF MICROORGANISMS
ISBN 0 12 410350 2

(1)   ease of use;
(2)   ease of storage—30 or 40 discs can be kept in a 14 mm screw-capped vial;
(3)   freedom from contamination—as the discs are kept dry there is no opportunity for growth of any contaminants introduced during sampling;
(4)   stability of characters—as the bacteria are not growing there is no opportunity for mutation and selection.

The method therefore has advantages over both active subculture on slopes and freeze-drying in ampoules.

II. DETAILS OF METHOD

The steps involved in preparing the discs are:

(1)   growth of bacteria;
(2)   suspension in gelatin;
(3)   distribution of drops to petri dishes;
(4)   freezing the drops;
(5)   freeze-drying the drops;
(6)   distribution of dried discs to vials;
(7)   checking the product.

1. Batch Size

The batch size prepared will depend on the intended use and distribution of the discs. As a guide, the base of one 9 cm petri dish will accommodate about 80 discs. This number should be more than enough for a year's supply for the average user. As a guide to scaling up the operation, an Edwards EFO3 freeze-dryer will accommodate a batch size of about 5000 discs dried in petri dishes. All volumes given in this method are for the single petri dish load of about 80 discs.

2. Growth of Bacteria

Bacteria may be grown on any suitable non-selective media. Nutrient or blood agar will be suitable for many strains. A single 150 × 19 mm tube of sloped medium will provide adequate growth of bacteria such as Enterobacteriaceae or staphylococci.

3. Harvesting

Growth is harvested with a Pasteur pipette in a minimal volume (about 0.5 ml) of nutrient broth and added to 3 ml of the gelatin suspending medium previously melted and held at 37°C. This is mixed well to suspend.

4. Gelatin-suspending Medium

Gelatin powder	(Oxoid)	10 g
Nutrient broth powder, No. 2	(Oxoid)	2.5 g
meso-Inositol	(Koch-Light)	5 g
Deionized water		100 ml

Dissolve the solids by gentle heating, check the pH and adjust if necessary to 7.2. Distribute to screw-capped 6 ml bottles (bijoux) in 3 ml volumes and sterilize by autoclaving at 121°C for 15 min.

5. Distribution of Drops and Freezing

With a dropping pipette delivering 0.02 ml ("fifty dropper"), drops of the suspension are placed in the base of a plastic petri dish. Maximum use of the area can be obtained by adopting a spiral pattern, starting from the outside. With care, about 80 drops can be accommodated in the base. The base of the petri dish is covered with the lid and carefully placed in a deep-freeze at −20 to −40°C until the drops are frozen. This is indicated by a change in appearance from transparent to opaque. Freezing will occur in about 20 min with a light load, but may take up to 2 h with a large batch. Drops freeze fastest in the petri dish at the bottom of the stack and dishes should be periodically rotated throughout the stack.

6. Freeze-drying

The petri dishes are transferred quickly to the freeze-dryer. If necessary, three piles can be accommodated under the plastic dome of the Edwards EFO3. The desiccant trays of the dryer must previously have been loaded with phosphorus pentoxide. The freeze-dryer is switched on and the cultures are dried overnight. With large numbers of discs in a batch it may be necessary to replace the phosphorus pentoxide after 2–4 h. This may be done by isolating the drying chamber, switching off the machine, and venting the trap before replacing the P_2O_5.

7. Preparation of Vials

Coarse, self-indicating silica gel (BDH Chemicals Ltd) is placed in 14 × 45 mm screw-necked vials (FBG Trident Ltd.) to a depth of about 10 mm and packed down tightly with a wad of cotton wool. The vials are sterilized in a covered container in a hot-air oven at 160°C for 1 h. The caps are sterilized separately by autoclaving at 121°C for 30 min. Caps are dried in an oven at 60–80°C for 4 h before placing on bottles.

8. Distribution of Discs to Vials

With small batches, discs may be transferred to vials with a small spatula. During drying the discs become detached from the plastic surface of the petri dish and are easily picked up with a spatula. When a large batch is prepared discs can be transferred to vials with a small, wide-necked funnel. Normal clean technique should be observed in distribution; strict asepsis is not necessary since any air-borne contaminants will have no opportunity to multiply on the dried discs. After distribution, the caps of the vials should be replaced and tightened.

9. Revival of Dried Bacteria

With fine-nosed forceps one gelatin disc is placed in 1 ml of nutrient broth. The broth is warmed in a 37°C incubator until the disc dissolves. A loopful of the broth suspension is transferred to a suitable solid medium and streaked to obtain single colonies before incubating.

10. Checking the Product

Bacteria preserved by this method must be checked for viability and for retention of the particular characteristics for which they have been preserved. As this method is unlikely to be used as a sole means of preserving important cultures, it is probably not necessary to perform viable counts on the discs as simple plating will give a good indication of the level of viability. As in any method of preservation, it is essential to characterize the strain after drying to ensure that the correct strain has been preserved and has retained its important characteristics.

III. STORAGE CONDITIONS

Discs are stored at 5°C. It is important that the vials are allowed to warm to room temperature before opening to prevent condensation of water in the vial. No comparative studies have been performed on the effect of different temperatures of storage but no drop in viability has been noted after short periods at room temperature. Obara *et al.* (1981) stored gelatin discs at −20°C.

IV. GENERAL NOTES

(1) Vented petri dishes should be used to allow escape of water vapour during drying.

(2) In warm weather the gelatin drops may not set after dropping and will coalesce when moved. This may be overcome by placing the petri dish on a layer of ice during the dispensing of the drops.

(3) If large batches are to be prepared drops can be placed in the base and the inside of the lid of the petri dish. The drops in the lid are frozen before replacing the lid on the petri dish.

(4) Distribution of the discs to vials is greatly facilitated if the discs have become detached from the surface of the petri dish. For unknown reasons, discs in petri dishes placed on the base plate of the freeze-dryer sometimes remain attached to the petri dish. A dummy layer of empty petri dishes on the base plate solves this problem.

V. SUCCESSFULLY PRESERVED ORGANISMS: SHELF LIFE

Various species of Enterobacteriaceae, staphylococci, *Pseudomonas aeruginosa* and *Corynebacterium diphtheriae* have been successfully preserved for at least 4 years. The method has not been successful with more delicate species such as *Neisseria* or *Haemophilus*. However, Obara *et al.* (1981) using a method described by Yamai *et al.* (1979) have reported successful preservation of *Neisseria*, *Haemophilus*, and *Bacteroides* using a gelatin-disc method based on a different suspending mixture.

7

Maintenance of Anaerobic Bacteria

C. S. IMPEY
Food Research Institute
Norwich, UK
and
B. A. PHILLIPS
National Collection of Dairy Organisms
National Institute for Research in Dairying
Reading, UK

I. INTRODUCTION

Although media and methods for the isolation and culturing of anaerobes have improved greatly over the past few years, the preservation of cultures still causes problems. Maintenance by periodic subculture is inconvenient, time-consuming and often results in contamination or loss of viability.

When cultures are freeze-dried the resulting ampoules, if correctly stored, will remain viable for periods far beyond any practical

MAINTENANCE OF MICROORGANISMS
ISBN 0 12 410350 2

requirement. In addition, the freeze-dried culture is the only practical form in which strains can be exchanged between researchers.

II. CULTURE METHODS FOR ANAEROBES

Current methods for the cultivation of anaerobes may be conveniently divided into those in which the media are prepared and used under an oxygen-free atmosphere, and those in which the media are prepared and inoculated in air and which may or may not be incubated with an oxygen-free headspace.

For strict anaerobes the method first described by Hungate (1950) may be used. All manipulations, including media preparation, are carried out under a constant flow of carbon dioxide or nitrogen which has been passed over heated copper to remove any remaining traces of oxygen. Further details of this method can be found in the publications of Hungate (1966, 1969), Latham and Sharpe (1971), Barnes and Impey (1974) and Holdeman et al. (1977). Anaerobic cabinets also maintain a constant oxygen-free headspace over the media and are now available from manufacturers of scientific equipment (Don Whitley Scientific Ltd; Raven Scientific Ltd).

However, many anaerobes can be grown without the use of the Hungate technique or anaerobic chambers provided conditions are properly controlled. This is especially true of clinical isolates and evidence now suggests that pathogenic anaerobes are among the more oxygen tolerant (Rosenblatt et al., 1973). Broth media containing reducing agents such as cysteine, together with small amounts of agar to minimize diffusion of oxygen, are used in 1-oz McCartney bottles with metal screw-caps. The bottles are filled with about 20 ml of broth to leave a minimal headspace and immediately prior to inoculation are held in a boiling water bath for 20 min to expel oxygen. After inoculation, with c. 0.25 ml of culture, the caps are screwed down tightly to prevent access of oxygen and the broths are then incubated at 37°C for 1–2 days. Agar plates should be pre-reduced in anaerobic jars for at least 24 h before inoculation and then returned to the jars for incubation. The gas mixture in the anaerobic jars is 80% N_2 + 10% H_2 + 10% CO_2; alternatively, gas-generating sachets (Oxoid Ltd; B-D Laboratory Products) can be used.

III. MEDIA

A. *Media for Use with the Hungate Technique*

1. MM10 broth modified from M10 (Caldwell and Bryant, 1966)

	Per litre
Cellobiose	1.0 g
Maltose	1.0 g
Glucose	1.0 g
Starch (soluble)	1.0 g
Yeast extract, B127 (Difco)	2.0 g
Trypticase, BBL11921 (B-D)	2.0 g
Mineral solution I	75 ml
Mineral solution II	75 ml
Haemin solution	10 ml
VFA mixture	3.1 ml
Resazurin solution	1.0 ml

pH 6.8

Separate additions: 20 ml/litre of 2.5% (w/v) L-cysteine hydrochloride
50 ml/litre of 8% (w/v) sodium carbonate

Preparation: All the ingredients except cysteine and carbonate are dissolved in glass-distilled water, the pH is adjusted to 6.8 with M NaOH and made up to 93% of final volume. This is sterilized in a cotton-wool plugged flask by autoclaving at 121°C for 15 min. The 2.5% (w/v) cysteine solution is prepared in a screw-cap McCartney bottle with a small headspace and the 8% (w/v) sodium carbonate solution in a small cotton-wool plugged flask with a large headspace; both are autoclaved at 121°C for 15 min. While the flasks are still hot from the autoclave, short, sterile gassing jets are inserted and oxygen-free carbon dioxide is passed over the surface of the medium and of the carbonate. When the medium has cooled to below 50°C the cysteine solution is added, followed by the carbonate; long gassing jets are inserted below the liquid surface and bubbled for at least ½ h. The complete medium is distributed into sterile stoppered tubes using standard Hungate technique and incubated at least overnight to check for contamination and for oxidized (pink) tubes. This method of preparation is that of Latham and Sharpe (1971) but an alternative method for preparing the medium is described by Barnes and Impey (1974).

(a) MM10 medium solutions and mixtures
   (i) Mineral solution I
       K_2HPO_4 6.0 g/litre
       in glass distilled water; store at 4°C
   (ii) Mineral solution II

	g/litre
NaCl	1.2
$(NH_4)_2SO_4$	12.0
KH_2PO_4	6.0
$CaCl_2$ (anhydrous)	1.2
$MgSO_4$ $7H_2O$	2.5

   in glass distilled water; store at 4°C
   (iii) Haemin solution
       10 mg of haemin is dissolved in 1 ml M NaOH and the
volume made up to 100 ml with glass-distilled water; store at 4°C.
   (iv) VFA Mixture

Acetic acid	17 ml
Propionic acid	6 ml
n-Butyric acid	4 ml
iso-Butyric acid	1 ml
DL-α-methyl-*n*-butyric acid	1 ml
n-Valeric acid	1 ml
iso-Valeric acid	1 ml
Phenyl-acetic acid	1 g

   Mix well and store at 4°C
   (v) Resazurin solution.
       0.1% (w/v) resazurin in sterile glass distilled water. Store at
       4°C.

(b) Additions and supplements for MM10
   (i) Agar—for agar slants 1.5% (w/v) agar (Difco) is added to the
       broth medium.
   (ii) Lactate—the growth of *Veillonella* and *Megasphaera* species is
       enhanced by the addition of sodium lactate to a final
       concentration of 1% (w/v).
   (iii) Rumen fluid—although this medium was developed to
       obviate the need for rumen fluid, some organisms such as
       Eubacterium cellulosolvens will grow much better if clarified
       rumen fluid (Bryant and Robinson, 1961) is added to 15%
       (v/v) final concentration.
   (iv) Faecal extract—Barnes and Impey (1978) found that a
       significant number of anaerobes of poultry origin had a
       requirement for an aqueous extract of chicken faeces. The

extract is prepared by autoclaving at 121°C for 30 min equal quantities (w/w) of chicken faeces and water. The sludge is centrifuged (1500 g for 20 min) and, after decanting, the supernatant is left at 1°C overnight. The supernatant is further centrifuged (16000 g for 20 min) to remove remaining debris and the pH is adjusted to 7.0–7.2. It is stored sterile (autoclave 121°C for 15 min) and when required is incorporated in the medium at a final concentration of 10% (v/v). Extracts of faecal material from other sources can be prepared and used in the same manner.

(v) Liver extract—Barnes and Impey (1978) also found anaerobes with a requirement for liver extract. This extract is prepared by dissolving 27 g of dehydrated liver, B133 (Difco Laboratories Ltd), in 200 ml glass-distilled water, heating to 50°C and holding at this temperature for 1 h. The mixture is then boiled, cooled, and centrifuged (1500 g for 20 min) and the pH of the supernatant adjusted to 7.0–7.2. The extract is stored sterile (autoclave 121°C for 15 min) and used in the medium at a final concentration of 5% (v/v).

B. *Media for Use in Screw-cap McCartney Bottles*

1. *VL broth (Barnes and Impey, 1971). Modified from Beerens et al.* (1963)

This contains (g/litre): tryptone, L42 (Oxoid Ltd), 10; NaCl, 5; beef extract, Lab-Lemco powder L29 (Oxoid Ltd), 2.4; yeast extract, B127 (Difco Laboratories Ltd), 5; cysteine hydrochloride, 0.4; glucose, 2.5; agar (New Zealand), 0.6; pH 7.2–7.4. The medium is sterilized by autoclaving at 121°C for 15 min in 19 ml amounts in 1 oz McCartney bottles with metal caps.

2. BGP broth

This is VL broth with the glucose level reduced to 1 g/litre and with the addition of Na_2HPO_4 at 4 g/litre.

3. VLhlf and BGPhlf

These media are VL and BGP broths containing 1 μg ml^{-1} haemin (see section A.1(a)(iii)), 5% liver extract (see section A.1(b)(v)) and 5% chicken faecal extract (see section A.1(b)(iv)).

4. Cooked meat broth

This can be made using either of the formulations described by Cowan (1974) or by using meat granules, Lab 24 (London Analytical and Bacteriological Media Ltd) in nutrient broth. The medium is distributed in 20 ml amounts in 1 oz screw-cap bottles and sterilized by autoclaving at 120°C for 15 min. It is recommended that the depth of meat granules should be about 5 cm.

IV. MAINTENANCE BY SUBCULTURE

A. Cultures Requiring the Hungate Technique

MM10 agar with supplements as required is used in tubes that are sloped with a deep butt. The medium is inoculated by stabbing into the butt. Cultures can be kept in this medium for 1–4 weeks and are stored at room temperature in the dark.

B. Cultures in Screw-cap Bottles

Anaerobes may be kept for 1–4 weeks in low glucose buffered media such as BGP or BGPhlf. The broths should be stored at room temperature in the dark with the caps screwed down tightly and only opened for use once. Non-sporing anaerobes are very sensitive to acid conditions and the use of unbuffered glucose containing media such as VL is not recommended.

Media suitable for non-sporing anaerobes may also be used for clostridia, but cooked meat broth is often more satisfactory. The more putrefactive readily sporing clostridia will survive in this medium for several years, but the poorer sporeformers for only 6–12 months.

V. MAINTENANCE BY FREEZING

Anaerobes that grow well on MM10 agar slopes can be stored at $-80°C$ for about 3 months. There are, however, a number of disadvantages with this method of preservation: the tube size limits the numbers that can be stored; the bungs used in the Hungate tubes harden in the cold and may become dislodged; and slow thawing may well damage cultures irreparably.

VI. MAINTENANCE BY FREEZE-DRYING

A. Apparatus and Equipment

1. Freeze-drying machines

Any centrifugal freeze-dryer is suitable for the preservation of anaerobic bacteria. The methods described here have been used in conjunction with freeze-dryer models 2A/110, 5PS, EF03 and EF6 (Edwards High Vacuum; A. D. Wood (London) Ltd).

2. Ampoules

Before use neutral glass ampoules (Adelphi Manufacturing; Glass Wholesale Supplies Ltd) are washed in detergent and thoroughly rinsed, first with tap and then demineralized water. They can be labelled by inserting strips of Whatman No. 1 filter paper (Scientific Supplies Co. Ltd) with the culture number and date of drying written either in pencil or non-toxic ink (Lapage *et al.*, 1970).

The ampoules are capped with 10 × 50 mm glass specimen tubes (Scientific Supplies Co. Ltd) or non-absorbent cotton-wool plugs and dry-sterilized.

3. Primary drying caps

Caps for primary drying are made using surgical lint sewn at the edge. The overall size is approximately 20 × 40 mm: the cap should be an easy fit on the ampoule, but tight enough to withstand handling and spinning in the dryer centrifuge. The caps are sterilized in glass Petri dishes.

4. Secondary drying plugs

Plugs for secondary drying are made by covering butter muslin with a thin (3–5 mm) layer of non-absorbent cotton wool and cutting into 12 mm squares. The squares are packed in glass petri-dishes, muslin side up (layers interleaved with paper), and dry-sterilized. Alternatively, plugs from sterile ampoules can be used.

5. Pipettes

Both standard 1 ml pipettes and 30 dropper Pasteur pipettes (Harshaw Chemicals Ltd; John Poulten Ltd) are plugged with non-absorbent cotton wool and dry-sterilized in cans. It is important that the 30 dropper pipettes are long enough to reach the bottom of the ampoules (*c.* 100 mm).

B. *Preparation of the Suspension for Freeze-drying*

1. Growth media

Organisms for freeze-drying must be grown in a medium giving rapid and vigorous growth. For strains grown using the Hungate technique, MM10 broth should be used and VL broth for screw-cap bottle cultures. Both media have supplements of rumen fluid, faecal extract, liver extract etc., added as required.

2. The suspension

Hungate cultures are centrifuged (2500 g for 15 min), the supernatant poured off, and 1 ml of suspending fluid added to the centrifuged deposit. The suspending fluid is 7.5% glucose in horse serum (Tissue Culture Services), sterilized by filtration (Phillips *et al.*, 1975). Up to the point where the supernatant is poured off, Hungate cultures should be kept under a carbon dioxide headspace, but once preparation of the suspension is complete this can be dispensed with provided there is no undue delay before freeze-drying is started.

Cultures in screw-cap bottles are centrifuged (2500 g for 15 min) and after the supernatant has been poured off 1 ml of 16% glucose, which has been held in a boiling water bath for 10 min and cooled just before use, is added to the centrifuged deposit.

C. Freeze-drying Method

1. Primary drying

The bacterial suspension is distributed in approximately 0.1 ml quantities per ampoule, the glass cap/cotton-wool plug is replaced by a lint cap and the ampoules placed in the dryer centrifuge. The time taken from filling the first ampoule to starting the freeze-dryer should be kept as short as possible—about 30 min for cultures grown in screw-cap bottles and 15 min for Hungate cultures. The centrifuge should be run for up to 10 min and primary drying continued for 3 h.

Freeze-dryers using phosphorus pentoxide, methanol/solid CO_2 cold fingers and mechanically frozen trays have all proved successful for drying anaerobes.

2. Ampoule constriction and secondary drying

The ampoules are removed from the primary drying chamber and the lint caps replaced with plugs. Using sterile pointed forceps the muslin/cotton-wool plugs are held in the centre and then pushed into the ampoule neck. If plugs from sterile ampoules are used they should be carefully transferred to the ampoule containing the dried material and then trimmed off to the required length (*c.* 10 mm). The plugs are pushed down the ampoule, using a rod fitted with a stop, so that the bottom of the plug is approximately 10 mm above the top of the dried material. The pointed forceps and the plug push rod should be sterilized by dipping in 70% alcohol and flaming between batches of ampoules of any one strain.

The ampoules are then constricted using either a constricting machine (Edwards High Vacuum; A. D. Wood (London) Ltd) or an oxy-gas torch (Jencons (Scientific) Ltd) with a fine jet. Care should be taken to make the bore of the constriction fairly fine but with thick walls, so that when the ampoules are finally sealed off a good strong point is formed.

Fit the ampoules onto the secondary drying head, fitting any unused nipples with empty ampoules. Where phosphorus pentoxide was used as the primary drying desiccant the trays are topped up with fresh material for secondary drying. With methanol/CO_2 cold fingers and mechanically refrigerated traps, phosphorus pentoxide in trays is also used as a secondary drying desiccant. Secondary drying (under vacuum) is continued overnight (15–20 h).

The ampoules are then sealed under vacuum using a Flaminaire torch (Longs Ltd) and tested for vacuum using a high-frequency vacuum tester (Edwards High Vacuum; A. D. Wood (London) Ltd). The discharge should be a blue-white colour; if the colour is mauve the ampoule should be left for a few hours and then retested. No discharge means no vacuum.

D. Opening Ampoules

The ampoule is marked with a glass knife at the level of the centre of the cotton-wool plug. The area of the mark is wiped with 70% alcohol, flamed, and then cracked by applying a fine, red-hot rod. Approximately 30 s is allowed for air to enter the ampoule, then the pointed end of the ampoule is removed. The cotton wool plug is replaced with a fresh one from a sterile ampoule. The dried culture material is reconstituted using suitable broth media; the medium is usually the same as that used to grow the organism for drying.

E. Quality Control

1. Viability
To check that the organism is still viable, one ampoule of each batch of a strain should be opened into a suitable broth.

2. Counting
In general, a simple growth or no-growth check for viability will be sufficient but, if an indication of survival rate is required, some counting method must be used. Surface plating or roll tubes can be used, but a much more practical method is to prepare tenfold dilutions to extinction using appropriate broth media (Phillips *et al.*, 1975).

3. Purity
Anaerobes, especially those cultivated by the Hungate technique, are very prone to contamination. When the test for viability is

carried out the culture should also be checked for contaminating aerobes and anaerobes. It is also useful to check the purity of the suspension used for drying.

F. Survival on Drying

Most anaerobes seem to be able to survive freeze-drying, but the loss in viable count will vary considerably. The variation is not dependent on genus; even different strains of the same species show varying survival levels. Also, the degree of anaerobiosis required to culture an organism is no index of its ability to survive freeze-drying; some of the strictest anaerobes show comparatively small losses on drying (Impey and Lee, unpublished data).

G. Shelf Life

Although the initial loss on drying may be high, if the ampoules are stored in the dark at a cool (2–10°C), even temperature further losses are low. The viability of anaerobes preserved in this way has remained high for at least 10 years, and even after 22 years cultures are still alive.

VII. GENERAL NOTES

(1) Rather than use a multiplicity of media for drying anaerobes it is often possible to use one medium for cultures in screw-cap bottles and one for cultures requiring the use of the Hungate technique provided the media selected will grow the most nutritionally demanding of the organisms being handled.

(2) Freeze-dryers require little maintenance, apart from the need to change vacuum pump oil and clean spilled phosphorous pentoxide from desiccant chambers. However, there are a number of seals that are made and broken each time the dryer is used and these will require replacement at fairly frequent intervals if the dryer is to maintain its performance. These seals include the secondary dryer nipples and the "O" and "L" rings of the desiccant chambers and centrifuge chamber bell jars.

(3) In order to minimize the exposure of cultures to oxygen the freeze-dryer should be prepared with all desiccant trays filled or mechanical freezers run down to the required temperature and the general vacuum tightness of the dryer checked before any suspensions are prepared for distribution into ampoules.

8

Maintenance of Leptospira

S. A. WAITKINS
Leptospira Reference Laboratory
Public Health Laboratory
Hereford, UK

I. INTRODUCTION

Leptospira are difficult to maintain for long periods; there is as yet no method which will guarantee a good recovery rate or preserve pathogenicity and antigenic pattern of the organisms. They may be kept alive, however, by time-consuming and tedious subculturing techniques or by storage in liquid nitrogen. All attempts at freeze-drying the organisms have been unsuccessful and earlier hopes of using this method have not been realized (Wolff, 1960; Otsuka and Manako, 1961 a, b; Annear, 1962; Resseler *et al.*, 1966; Coghlan *et al.*, 1967).

II. SUBCULTURING

The frequency of subculturing depends on the medium used, the incubation temperature, and the dimensions of the container. In general, the narrower the tube and deeper the medium, the greater is the likelihood of successful maintenance.

MAINTENANCE OF MICROORGANISMS
ISBN 0 12 410350 2

A. Glassware

It is important that all glassware should be perfectly clean and free from soap or detergent residues. The cleaned glass should be soaked in Dulbecco phosphate buffer solution at pH 7.6 and rinsed in distilled water.

B. Media

Liquid or semi-liquid medium is used. Semi-liquid medium contains 0.5% agar (w/v) and allows longer intervals between subculturing, usually up to 12 weeks. Two media are recommended for maintenance and are outlined below.

1. Korthof's medium (Korthof, 1932)

(a) Base A

Neopeptone (Difco)	0.8 g
NaCl	1.4 g
NaHCO$_3$	0.02 g
CaCl$_2$	0.04 g
KH$_2$PO$_4$	0.24 g
Na$_2$HPO$_4$.2H$_2$O	0.88 g
Distilled water	1 litre

(b) Base B (Inactivated rabbit serum). Rabbit serum that is negative for leptospiral agglutinens should be used. Blood is collected from an ear vein and allowed to clot; the serum is then pipetted off, inactivated at 56°C for 30 min, and sterilized by Seitz filtration.

(c) Preparation. The ingredients of Base A should be steamed at 100°C for 20 min, cooled, and filtered through double thickness Whatman No. 1 paper (Scientific Supplies Co. Ltd.) The pH is adjusted to approximately 7.2. The whole volume is dispensed in 100 ml amounts and autoclaved at 115°C for 15 min. When cool, 8 ml of previously sterilized rabbit serum (Base B) is added aseptically to Base A and mixed thoroughly. The medium is distributed into either 150 × 13 mm glass tubes or sterile bijoux bottles with plastic caps; both should be tightly closed.

The dispensed medium should be incubated first at 37°C for 2 days and then at 22°C for a further 3 days to ensure the sterility of the product.

If semi-solid medium is used, 5 g of agar is added to Base A. Just before use 2 to 3 drops of fresh blood from a sero-negative guinea pig are added to each container. When satisfactory growth has been achieved the tube is closed firmly with a rubber bung and stored either at 28°C or at room temperature (*c.* 22°C). Turner (1970) showed that using this medium leptospira survive up to 12 years.

2. Ellinghausen and McCullough (EM) medium (Ellinghausen and McCullough, 1965 a, b)

(a) Stock solutions. (The stock solutions are stored at −20°C until required.)

(i) Phosphate buffer, concentrated × 25

Na_2HPO_4	16.6 g
KH_2PO_4	2.172 g
Distilled water	1 litre

(ii) Salts, concentrated × 20

NaCl	38.5 g
NH_4Cl	5.35 g
$MgCl_2.6H_2O$	3.81 g
Distilled water	1 litre

(iii) Copper sulphate solution

$CuSO_4.5H_2O$ 30 mg/100 ml distilled water

(iv) Zinc sulphate solution

$ZnSO_4.7H_2O$ 80 mg/200 ml distilled water

(v) Ferrous sulphate solution

$FeSO_4.7H_2O$ 500 mg/200 ml distilled water

(vi) Vitamin B12

a. concentrate	10 mg/100 ml distilled water
b. working solution	10 ml concentrate + 90 ml distilled water

(vii) Vitamin B1   200 mg/100 ml distilled water

(b) Tween 80 (1% solution). The Tween 80 stock bottle (BDH Chemicals Ltd) is placed in a water bath at 56°C together with a 1 litre flask containing 480 ml of distilled water. When both have equilibrated to 56°C, 4.8 ml Tween 80 is slowly pipetted into the warmed distilled water, taking care to rinse all traces of Tween 80 from the pipette. The 1% solution is shaken gently to mix.

(c) Bovine albumin solution. (This solution should be freshly prepared before use.)

	1 litre
Pentax bovine albumin, Fraction V (Armour Pharmaceuticals)	50 g
Phosphate buffer concentrate (stock solution)	40 ml
Distilled water	960 ml

The distilled water is boiled and when cooled the concentrated phosphate buffer is added. The bovine albumin is gently dissolved in the prepared buffer solution. The pH is adjusted to 7.4 with 0.4N NaOH. The solution is sterilized using Seitz filtration pads (Baird and Tatlock (London) Ltd) or membrane filters (Millipore (UK) Ltd) (pore size 0.22μ).

(d) L-cystine (BDH Chemicals Ltd). 0.8 g.

(e) Inactivated rabbit serum. 50 ml. Prepared as in 1 (b).
It has been found that adding inactivated rabbit serum enables maintenance of leptospira for longer periods.

(f) Preparation. 2800 ml of distilled water is added to a 5 litre flask and the following solutions added: 160 ml of 25 × phosphate buffer stock solution (i); 200 ml of 20 × salts stock solution (ii); 4 ml copper sulphate stock solution (iii) and 40 ml of zinc sulphate stock solution (iv). 80 ml of the ferrous sulphate solution (v) is added and the mixture shaken for 5 min. The medium will become hazy at this point, probably due to precipitation of insoluble phosphate. L-cystine (d) (0.8 g) is added and the mixture shaken for a further 3 min. No attempt is made to completely dissolve the L-cystine. The mixture is then filtered through a triple thickness of Whatman No. 1 filter paper (Scientific Supplies Co. Ltd.) The resulting solution should be clear.

The previously prepared Tween 80 solution (b) is added and the whole thoroughly mixed before adding a further 236 ml of distilled water. This base bulk solution is now autoclaved at 121°C for 15 min and allowed to cool.

When the base bulk solution has cooled, 1 litre of freshly prepared sterile bovine albumin solution (c) is added aseptically together with 80 ml of vitamin B12 working solution (vi b), 0.4 ml of vitamin B1 solution (vii) and 50 ml of sterile rabbit serum (e). The pH is checked and readjusted, if necessary, to pH 7.3 to 7.4 The complete medium is then sterilized by filtering through a "sandwich" of membrane filters (pre-filter pad, 0.22μ, 0.45μ, and 1.2μ pore-size membranes) (Millipore Ltd) into either sterile glass tubes or bijoux bottles with plastic caps. The medium can also be dispensed into 100 ml amounts for use later.

The dispensed medium is then incubated at 37°C for two days and at 22°C for three days to ensure sterility.

Both media are very complicated to prepare, require considerable technical skill to manipulate and, in general, require sub-culture of the leptospira every 6 to 12 weeks.

C. Method
Isolates and stock cultures of strains for antigen preparation should

be subcultured in either Korthof's medium or Ellinghausen and McCullough's medium with added 0.5% agar (semi-solid). Cultures should be incubated at 28°C until good growth is obtained. They should be examined by dark-field microscopy (× 40) for possible contamination and viability. Only those cultures demonstrating actively motile leptospira should be subcultured into further maintenance medium; those that are growing poorly should be transferred to liquid medium containing an increased concentration of rabbit serum (Turner, 1970) and reincubated at 28°C until profuse growth is seen. The whole subculturing procedure should be repeated every 12 weeks.

III. LIQUID NITROGEN

By far the easiest method of maintaining leptospira is storage in the vapour phase of liquid nitrogen. At the Leptospira Reference Unit, Hereford, organisms suspended in E.M. medium have been recovered from liquid nitrogen after storage for up to 2 years. However, recovery rates are low, usually about 10–20%, and the optimal conditions for long-term survival have yet to be established. Nevertheless, preserving leptospira in liquid nitrogen has been found easy to carry out and eliminates the need for continuous subculturing. With immediate subculturing into fresh medium following thawing, many of the primary functions of leptospira, such as pathogenicity and antigenic pattern, have been preserved.

A. Method

Leptospira are grown in the Ellinghausen and McCullough medium described in section II B.2 until profuse growth is obtained. 0.5 ml of the culture is put into a 2 ml capacity plastic ampoule (Sterilin Instruments) and placed in the vapour phase of liquid nitrogen. When retrieval is required the ampoules are thawed quickly and the culture is immediately inoculated into Ellinghausen and McCullough medium supplemented with an increased volume of inactivated rabbit serum (section II.B.1.b) (20%). Subculture into fresh medium is usually required before full viability is re-established.

As mentioned previously, recovery rates using liquid nitrogen are low, usually about 10–20%.

IV. CONCLUSIONS

It is obvious that maintaining leptospira in a vital and active state is extremely difficult. Although some measure of success is possible using the methods described, there remains a desperate need to establish alternative methods for maintaining these organisms.

Future developments should re-examine freeze-drying, for only this method of maintenance ensures that leptospira organisms keep their primary pathogenic and antigenic function intact.

9

Maintenance of Industrial and Marine Bacteria and Bacteriophages

I. J. BOUSFIELD

National Collections of Industrial and Marine Bacteria Ltd
Torry Research Station
Aberdeen, UK

I. INTRODUCTION

Bacteria may be successfully maintained using a number of different methods. The methods described in this chapter have been used by the National Collections of Industrial and Marine Bacteria (NCIMB) for a wide range of species.

II. FREEZE-DRYING AND SUSPENDING MEDIA

Most species of bacteria in the NCIMB are maintained by a standard freeze-drying method essentially the same as that described by Rudge (this volume, ch. 4). A special feature of the NCIMB method is the use of "*mist. desiccans*" (Fry and Greaves, 1951) as a suspending medium.

"*Mist. desiccans*", which is a mixture of horse serum (Wellcome

MAINTENANCE OF MICROORGANISMS
ISBN 0 12 410350 2

Reagents Ltd, No. 3), nutrient broth (Oxoid Ltd, CM1) and glucose, has been found to be one of the best general purpose freeze-drying menstrua. It is effective for a very wide range of organisms, including some of the photosynthetic bacteria and iron-oxidizing sulphur bacteria which do not always freeze-dry well in other media. In our experience it is superior to simple mixtures of serum and carbohydrate (including serum and inositol) for the more sensitive organisms. Double-strength skim milk is an acceptable alternative in most cases, but survival of sulphate-reducing bacteria is not always good. Some years ago, in an attempt to find a freeze-drying medium more convenient to prepare than "*mist. desiccans*", experiments were carried out at the NCIMB with sucrose/sodium glutamate mixtures. Although initial survival levels were often high, the viability of some organisms (notably streptococci) declined substantially during storage. Therefore, sucrose/sodium glutamate mixtures are not recommended for the long-term storage of freeze-dried bacteria.

1. Preparation of "Mist. desiccans"

75 ml serum (Wellcome Reagents Ltd, CM1) and 25 ml nutrient broth (Oxoid Ltd, No. 3) are mixed in a 250 ml conical flask; 7.5 g glucose are added a little at a time, while the contents of the flask are swirled gently. When the glucose has dissolved, the mixture is sterilized by membrane or Seitz filtration using *pressure* filtration, not vacuum filtration which causes troublesome frothing. The sterilized mixture is dispensed aseptically in 5 ml amounts in sterile screw-capped bottles and incubated at 25–30°C for 5 days as a sterility check. Five ml "*mist. desiccans*" is used to harvest the growth from three universal bottle agar slope cultures. This will yield an adequately dense suspension which can be dispensed in 0.1 ml amounts into about 25 freeze-drying ampoules.

The rest of the freeze-drying procedure, storage and revival are essentially as carried out using the method described by Rudge (this volume, ch. 4).

III. L-DRYING

l-drying (Annear, 1958) is a useful method of vacuum-drying for the preservation of bacteria that are particularly sensitive to the initial freezing stage of the freeze-drying process. The essential feature of L-drying is that the cultures are not allowed to freeze, but are dried direct from the liquid phase.

1. Method

Ampoules are prepared with numbered, dated filter-paper strips and cotton-wool plugs as described by Rudge (this volume, ch. 4). A dense suspension of bacteria in *"mist. desiccans"* is prepared and 3 drops (c. 0.1 ml) are dispensed into the sterile ampoules with a sterile 30-dropper pipette. The cotton-wool plugs are trimmed and pushed down into the ampoules with a ramrod (Rudge, this volume, ch. 4, section II.7). The ampoules are attached vertically to the underside of a horizontal manifold, clamped above a glass tank of water at 20°C, and connected with a diaphragm valve and P_2O_5 trap to a rotary vacuum pump (Fig.1). The water level in the tank should be such that the ampoules are immersed to a depth of 40–50 mm.

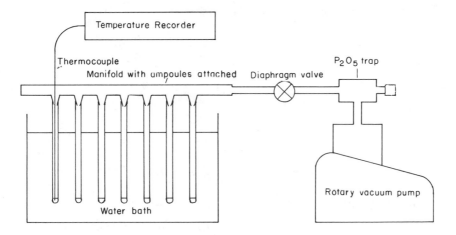

Fig. 1: Diagram of L-drying apparatus.

With the valve to the manifold closed, the pump is switched on. The valve is then opened wide for 0.5 s and quickly closed again, removing most of the air from the system without causing violent removal of dissolved air from the suspensions. The valve is then opened again very gradually until the contents of the ampoules start to bubble gently. The rate of bubbling is controlled by careful manipulation of the valve so that the contents of the ampoules degas without violent bubbling. If this occurs the valve should be closed quickly, to allow the bubbles to collapse, and then reopened cautiously. Degassing usually takes about 5 min, after which time the valve can be opened further to allow drying to take place. It is unlikely that the suspensions will freeze at this stage and, when there is no longer any danger of frothing, the valve can be opened fully.

After *c.* 30 min, the contents of the ampoules should appear to be dry. The ampoules are removed from the manifold, constricted, attached to a secondary-stage dryer, dried overnight, and sealed off as described in the section on freeze-drying (Rudge, this volume ch. 4). Cultures are resuscitated in the same way as freeze-dried cultures.

2. Shelf Life

In the NCIMB spirilla and *Azomonas insignis* are preserved by L-drying. These organisms are particularly sensitive to freeze-drying, but L-dried cultures have survived with good recovery levels for 10 years. It seems likely that they will survive even longer but, at present, experimental evidence for this is unavailable.

IV. MICRODRYING

Microdrying is a modification of the freeze-drying method used at NCIMB. It involves fewer manipulations than conventional centrifugal freeze-drying, resuscitation requires less skill, and there is a lower risk of contamination as the resuspending stage is omitted.

1. Method

Ampoules are prepared exactly as for freeze-drying except that the filter-paper strip is a thick grade (Genzyme Biochemicals Ltd, no. 17). Bacterial suspensions, as dense as possible, are prepared in *"mist. desiccans"* (see section II.1). Three drops (0.1 ml) are dispensed from a 30-dropper pipette on the filter paper strip in the ampoules. The pipette and ampoules should be held almost horizontally so that the suspension falls only onto the filter paper and not onto the wall of the ampoule. The filter paper should absorb all the suspension so that it appears damp without there being any excess liquid. Cultures can then be freeze-dried in the usual way, except that centrifugation and the replacement of cotton-wool plugs with lint caps are not necessary.

Resuscitation of microdried cultures is straightforward and does not require any manipulations with pipettes. The ampoule is first shaken or tapped to ensure that the filter paper strip is loose. The ampoule is then opened (Rudge, this volume, ch. 4, section II.12), the plug is removed and the dry filter-paper strip tipped aseptically into a broth or onto a slope of the appropriate growth medium.

2. Shelf Life

Microdrying has only recently been introduced in the NCIMB and thus there are no data available on long-term survival. However, the wide range of bacteria tested so far have shown immediate post-drying recovery levels comparable with those obtained with conventional freeze-drying, and there is no reason to suspect that long-term survival levels should not be similarly comparable.

V. FREEZING OVER LIQUID NITROGEN

Freezing over liquid nitrogen may be used for the preservation of a wide range of bacteria. At the NCIMB the method is used for microorganisms which do not survive freeze-drying or L-drying, for patent deposits, sensitive mutants, genetically manipulated strains and all bacteriophages. These last are preserved by freezing over liquid nitrogen, not because they cannot be freeze-dried or L-dried (although some cannot), but because of the risk of contaminating freeze-drying equipment, and hence other cultures, with phage particles.

1. Method for Bacteria

Polypropylene 2 ml cryotubes (Gibco-Europe Ltd) containing about 20 3 mm glass beads (available from most craft shops) are wrapped individually in greaseproof paper, sterilized by autoclaving at 121°C for 15 min and dried at 50°C.

Bacteria are grown up as for freeze-drying (Rudge, this volume, ch. 4, section II.1) and are thickly suspended in a suitable medium containing 10% (v/v) glycerol as a cryoprotectant. The suspension is pipetted into 2 cryotubes so that the glass beads are totally immersed. The contents of the tubes are agitated gently with the pipette to ensure that the beads are thoroughly coated with the suspension, after which excess liquid is pipetted off. The tubes are capped and placed in aluminium storage racks (Union Carbide) which are then placed in an LR 40 liquid nitrogen refrigerator (Union Carbide). The racking arrangement is such that the tubes are kept above the surface of the liquid nitrogen in the vapour phase to prevent seepage of liquid nitrogen into the tubes, with possible explosion on rapid thawing.

Cultures are resuscitated by chipping a glass bead off the frozen mass with sterile forceps and dropping it into appropriate growth medium.

2. Shelf Life of Bacteria

The NCIMB has been using the glass-bead method for about 3 years only and, therefore, has limited information on shelf life. However, with the exception of certain obligate methylotrophic bacteria, for which an ideal preservation method has yet to be found, survival appears to be good.

3. Method for Bacteriophages

High titre phage lysates are prepared using media and methods appropriate to the phage to be preserved. Host cells and debris are removed by low speed centrifugation (*c.* 3000 g) followed by membrane filtration (pore size 0.45 μm). The cell-free lysates are then dispensed aseptically in 1 ml amounts in previously sterilized 2 ml cryotubes (Gibco-Europe Ltd). The tubes are capped, then frozen and stored in a liquid nitrogen refrigerator as described in 1. above.

4. Shelf life of Bacteriophages

The NCIMB has evidence that, after an initial loss of titre of *c.* one order of magnitude, most coliphages will survive without significant loss for at least 7 years. However, phage MS2 does not survive for longer than *c.* 1.5 years. Marine phages also tend to have shorter shelf lives, the most notable example being the PM2 *Pseudomonas* phage, which is unlikely to survive more than 2–3 years.

10

Maintenance of Methanogenic Bacteria

H. HIPPE

Deutsche Sammlung von Mikroorganismen
Gesellschaft für Biotechnologische Forschung mbH.
Göttingen, Germany

I. INTRODUCTION

Methanogenic bacteria form a diverse group of procaryotic organisms physiologically united by the ability to produce methane as a result of their energy metabolism. They differ from the classical bacteria in such characteristics as their cell-wall polymers, membrane lipids, sensitivity to antibiotics, unique cofactors and sequence of 16S rRNA; they are known as archaebacteria (Balch *et al.*, 1979; Hilpert *et al.*, 1981; Kandler, 1982).

Methanogenic bacteria are obligate anaerobes; most of which are very sensitive to oxygen even when briefly exposed. Spores, or other resistant structures which could otherwise help survival under adverse conditions, are not known.

Currently, 27 species belonging to 11 genera have been described. With few exceptions, all methanogenic bacteria grow on a $H_2 + CO_2$ gas mixture; in addition, most of them utilize formate and some grow on acetate, methanol, or methylamines. The optimum temperature for growth of the various methanogens covers a remarkably broad

MAINTENANCE OF MICROORGANISMS
ISBN 0 12 410350 2

range from 20–25°C up to 85°C. The nutritional requirements are diverse; some strains are able to grow autotrophically in simple mineral media, whereas others require vitamins, cysteine, branched chain fatty acids, coenzyme M, acetate or unknown growth factors present in certain peptones or in rumen fluid.

Three media have been described which fulfil the requirements of many of the existing methanogenic pure cultures (Balch *et al.*, 1979). The preparation of media, as well as the successful cultivation of these bacteria, depends on the use of special equipment and techniques (Hungate, 1950, 1969; Macy *et al.*, 1972; Miller and Wolin, 1974; Balch and Wolfe, 1976; Zeikus, 1977). Three types of culture vessels are used for growing methanogenic bacteria in small volumes: the screw-capped anaerobic culture tubes described by Hungate (Bellco Glass Inc., 2047–16125); the Balch serum tubes (Bellco Glass Inc., 2048–18150); and serum bottles. Both the latter are sealed with butyl rubber septa or stoppers held in place with crimped aluminium seals (Wheaton Scientific Div., 224183 and 224303). They are especially convenient for the growth of methanogens in a pressurized atmosphere of H_2+CO_2 at 200 kPa which reduces the frequency with which the consumed gas mixture must be replaced (Balch and Wolfe, 1976; Balch *et al.*, 1979). The screw-capped "Hungate tube" can be pressurized to at least 100 kPa overpressure.

"Hungate tubes" containing 3–4 ml of agarized medium are also convenient for preparing dilution series in roll tubes, which are used for viability determinations by colony counts.

Agar plates can be used if a reliable anaerobic glove box is available. Plates are poured and inoculated inside the box, placed in a stainless steel anaerobic jar and removed from the box for incubation. Such anaerobic jars are equipped with a manometer and valves and can be pressurized with oxygen-free gas mixtures (Balch *et al.*, 1979).

However, some methanogens appear to be unable to grow in or on the surface of agar media and polysilicate plates have been developed for isolating these strains (Stetter *et al.*, 1981; Wildgruber *et al.*, 1982). The ability of pure strains to grow on different kinds of agar has not been tested.

Methods for maintaining cultures of the methanogenic bacteria may be grouped into: (i) subculturing, (ii) freezing, and (iii) freeze-drying. As the majority of methanogenic cultures have been isolated comparatively recently, experience is still limited, particularly with respect to their long-term preservation.

II. SUBCULTURING

In general the viability of most methanogenic bacteria can be prolonged when actively growing cultures are removed from the appropriate incubation temperature and kept at room temperature or refrigerated at 4°C. In this respect there is no major difference from non-archaebacterial anaerobes.

Cultures stored at reduced temperature remain viable for some weeks, provided no oxidation of the medium occurs. Species of the rod-shaped genera *Methanobacterium* and *Methanobrevibacter*, which are usually grown with a H_2+CO_2 gas mixture, appear generally more resistant than species of other genera (Zeikus and Wolfe, 1972; Zeikus and Henning, 1975; Zehnder and Wuhrmann, 1977). There is no information available on survival times of formate-grown cultures.

In contrast to most other anaerobes, the methanogenic bacteria are grown either in media with relatively low organic nutrient content or in mineral media. Compared with rich media, simple media are not well protected against traces of oxygen, which may diffuse through the rubber septa or stoppers during storage of stock cultures. Pressurizing the culture vessels with the appropriate gas mixture before storing them at reduced temperature reduces the possibility of oxygen entering. Cultures of *Methanosarcina barkeri* grown on methanol or methylamines lose their activity after depletion of substrate, at a rate depending on the strain, and some need to be subcultured at intervals of 3–7 days. However, when grown with acetate or H_2+CO_2, they often remain viable for several weeks. *Methanococcus voltae* tends to lyse rapidly after growth at the optimal temperature of 37°C (Whitman *et al.*, 1982); *Mc. vannielii* requires weekly transfer when grown on formate (Jones and Stadtman, 1977); again, *Methanothermus fervidus*, a highly thermophilic methanogen with an optimal growth temperature of 85°C, dies within a few hours after exhaustion of the H_2+CO_2 gas atmosphere. The atmosphere must be renewed, therefore, or the culture transferred before the gas mixture is completely depleted (Stetter *et al.*, 1981). The following technique has been found useful for maintaining a number of methanogens grown in liquid culture with a H_2+CO_2 gas mixture (Stetter *et al.*, 1981).

A. Method

Fresh medium is inoculated with 2–5% (v/v) inoculum from an active culture. After an appropriate incubation time (*c.* 6 h for the fast-growing strains; overnight for the slow-growing strains) at the optimum temperature, the culture vessel is repressurized with the same gas mixture to 200 kPa and stored at 4°C. Such cultures may be viable for several months. Samples from these cultures are either used to inoculate fresh media or, when a larger volume of an active culture is required, the whole of the stored culture is directly incubated. The applicability of this storage technique has not yet been proved for all types of methanogens.

III. FREEZING

As with other bacteria, freezing and storage at low temperatures is the most simple and effective way to preserve methanogenic bacteria. Several strains have been found viable after more than 5 years following freezing with 50% glycerol as cryoprotectant and storage at −70°C (W. E. Balch, pers. com.). With this method equal volumes of mid-logarithmic cultures in 15 ml serum vials are mixed with sterile degassed and pre-reduced (0.05% cysteine plus 0.05% sodium sulphide) glycerol. After being pressurized to 200 kPa of nitrogen, the vials are placed in the freezer at −70°C.

The method of choice for the long-term preservation of cultures of methanogens is to freeze and store them in liquid nitrogen. Formate grown cultures of *Methanococcus vannielii* have been maintained by freezing in liquid nitrogen supplemented with sucrose to a 10 to 20% final concentration (Jones and Stadtman, 1977). Again, excellent results have been obtained at the Deutsche Sammlung von Mikroorganismen (DSM) with over 40 strains representing all existing species. The procedure used at the DSM is described below.

A. Freezing in Glass Ampoules

1. Preparation of cell suspension

Cultures are grown in heavy-walled, round-bottomed bottles (*c.* 70 ml volume) with necks that can be closed with a butyl rubber septum and a screw-cap as with Hungate anaerobe tubes (Bellco Glass Inc., 2047–16125). Such bottles can easily be made by a glassblower. They fit in a normal laboratory centrifuge and are used both for growing the cells and for centrifugation.

Cells are cultured with a $H_2 + CO_2$ gas mixture in 20 ml of medium per bottle. Methanol is used as a growth substrate in 50 ml of medium per bottle for strains of *Methanosarcina, Methanococcus mazei*, and *Methanolobus*.

Cultures are grown to an optical density of about 0.300 at 600 nm, by which time about 4 ml of gas per ml of culture has been consumed. Cultures that use methanol are harvested before active gas production ceases. The cultures are centrifuged directly in the unopened bottle, after which the bottle is opened, a gassing cannula inserted and the supernatant aseptically removed as completely as possible with a 20 ml hypodermic syringe and a long needle of 2 mm diameter. To the cell pellet, 2 ml of suspending medium is added and cells are suspended by means of a Pasteur pipette, bent and heat-sealed at the tip. Depending on the number of ampoules that have to be prepared, cell suspensions from several cultures are collected in one bottle by transferring with a sterile Pasteur pipette.

2. Suspending medium and cryoprotectant

Fresh culture medium containing 10% (v/v) glycerol or 5% (v/v) dimethyl sulphoxide (DMSO) is used as a suspending medium. It is prepared by adding 0.55 ml of sterile glycerol or 0.27 ml of sterile DMSO to 5 ml of sterile reduced fresh medium in a culture tube. The cryoprotectant is sterilized separately by autoclaving in anaerobe test tubes under nitrogen atmosphere and is reduced before use by adding appropriate amounts of cysteine and sodium sulphide.

3. Filling ampoules

Clean, sterile, dry glass ampoules, such as the 2 ml Wheaton gold-band cryule (Wheaton Scientific Div., 651486) are filled with 0.1 or 0.2 ml of cell suspension using a calibrated Pasteur pipette connected to a pipette aid depositing the suspension at the bottom of the ampoule. Care is taken not to touch the inner side of the ampoule. The pipette is flushed with oxygen-free gas before aspirating the cell suspension. Similarly, the ampoule is flushed during filling by inserting a gassing cannula near the bottom. A mixture of N_2+CO_2 is used instead of the H_2-containing gas mixture. After filling, the gassing cannula is moved near the top of the neck; then, while gassing is continued, the ampoule is heated in the middle of the long neck with a thin, hot flame from a gas burner, drawn out and sealed. Each ampoule prepared this way is immediately placed in an ice bath to prevent further warming of the contents from the heated upper part of the ampoule.

An amount of cell suspension equal to that used to fill the ampoules is transferred to 5 ml of fresh medium and used for the determination of pre-storage viability.

4. Freezing and storage

Methanogenic bacteria do not need to be frozen at a carefully controlled cooling rate if the suspensions are protected by glycerol or DMSO. Freezing, therefore, is simply achieved by placing the ampoules (dried outside after being removed from the ice bath) in the cold gas phase of the liquid nitrogen, using the storage system (aluminium canes or drawers) supplied with the refrigerator. This method provides a cooling rate of about 50°C min^{-1} and gives good survival for routine work. For safety, the glass ampoules should be stored only in the vapour phase and not immersed in liquid nitrogen.

5. Recovery and determination of viability

For recovery, an ampoule is removed from the liquid nitrogen (using protective gloves) and quickly thawed by immediately placing it in warm water at 30–37°C. The ampoule is agitated until thawing is complete. The ampoule is dried outside and scratched with a diamond pencil at the constriction. This area is wiped with ethanol and flamed. The ampoule is broken open and the suspension immediately aspirated in a 1 ml hypodermic syringe (flushed with oxygen-free gas) and injected through the rubber septum of an unopened culture tube containing 5 ml of fresh medium.

For determination of viability, decimal dilutions are prepared in a series of nine tubes using the syringe technique. A culture tube containing 3.6 ml of molten 1.5% agar medium kept at 45°C is inoculated from each dilution and agar roll tubes are prepared. Both the liquid dilution series and the roll tubes are incubated under appropriate conditions. Colonies grown in the roll tubes can easily be counted under a stereo microscope. A comparison of the counts before and after freezing is used to calculate the survival ratio. If an anaerobic glove box is available agar plates can be used instead of roll tubes.

Some methanogens, e.g. *Methanoplanus limicola* and *Methanothermus fervidus*, will not grow on the surface of, or in, agarized medium (Stetter *et al.*, 1981; Wildgruber *et al.*, 1982). Furthermore, the preparation of agar roll tubes may be laborious and time-consuming in routine work. In this case, only the liquid dilution series is made. After incubation, a rough estimation of the viability is made based on the last tube-yielding growth. With the slow-growing methanogens, the incubation period has to be extended to several weeks.

B. Freezing in Glass Capillary Tubes

The use of glass capillary tubes instead of glass ampoules in the freeze-preservation of bacteria and protozoa has been described by Pautrizcl and Carloz (1952), Cunningham *et al.* (1963), Walker (1966), and Jarvis *et al.* (1967). In combination with the "Hungate technique", the glass capillary method has proved to be equally convenient for preserving strict anaerobes like the methanogenic bacteria.

Glass capillary tubes have several advantages compared with other miniaturized methods such as the plastic straws or glass-bead techniques (Nagel and Kunz, 1972; Feltham *et al.*, 1978; Gilmour *et al.*, 1978). They can be sealed hermetically and, because of their impermeability to gases, anaerobes are well protected against oxidation despite the large surface/volume ratio of the enclosed cell suspension. For recovery, a single capillary can be removed without thawing or contaminating the parallel samples. Again, glass capillaries can be stored in the vapour phase of liquid nitrogen as well as in the liquid phase, and are heavy enough not to rise to the surface when immersed. However, a tight seal is very important if capillaries are to be stored in the liquid phase. As with glass ampoules, improperly sealed glass capillaries take up nitrogen and will explode on removal from the cold. However, a perfect seal is easier to achieve with capillaries than glass ampoules.

Figure 1 shows the procedure for the glass capillary method as used for methanogens and other bacteria at the DSM for nearly 10 years.

1. Method

The cell suspension is prepared as described in A.1 and 2 above. About 1 ml of the suspension is transferred into a small sterile vial which is kept anaerobic by gassing with the appropriate gas mixture. The vial is placed in an ice bath. One glass capillary (Difco Laboratories, 0658–33) is taken from the sterile stock by fitting it to the tip of a micropipetter (Rudolf Brand) and enough of the cell suspension is aspirated to fill one-third of the length of the capillary. The volume taken up is *c.* 0.025 ml. The suspension is aspirated until it is 1 cm from the free end of the capillary which is sealed in a fine, hot gas flame. The second seal is made *c.* 2.5 cm away from the other end (a mark is made before sterilizing the capillaries) by heating and drawing out.

As capillaries are prepared, they are placed in 75% ethanol or a disinfectant solution. All capillaries are examined for proper seals under a stereo microscope. Because moist capillaries will freeze

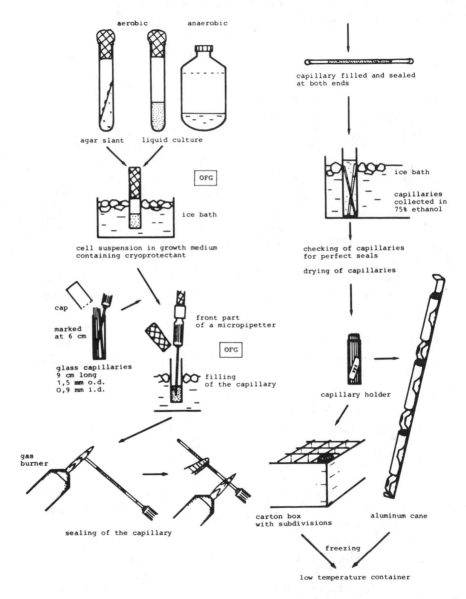

Fig. 1a: Glass capillary tube method for low-temperature preservation of microorganisms—filling and sealing of capillary tubes.

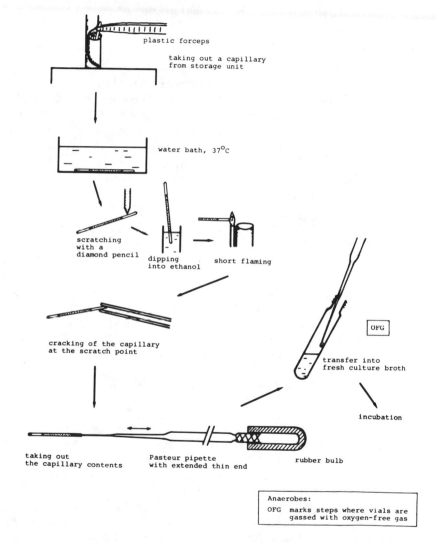

plastic forceps

taking out a capillary
from storage unit

water bath, 37°C

scratching
with a
diamond pencil

dipping
into ethanol

short flaming

cracking of the capillary
at the scratch point

OFG

transfer into
fresh culture broth

incubation

taking out
the capillary contents

Pasteur pipette
with extended thin end

rubber bulb

Anaerobes:
OFG marks steps where vials are
     gassed with oxygen-free gas

Fig. 1b: Glass capillary tube method for low-temperature preservation of microorganisms—removal of tubes from freezing storage, opening, and recovery of cell suspension.

together when cooled, they are dried by placing them between absorbent paper and gently pressing and rolling with the flat of the hand. The capillaries are placed in a small vial (quiver) that is labelled on the outside and left open. This is placed in the vapour phase of liquid nitrogen for freezing and storage.

Shock freezing by immersing capillaries directly in liquid nitrogen should be avoided because not all capillaries will withstand rapid cooling and some will break.

2. Recovery

A capillary is removed from the refrigerator and thawed rapidly in warm water. The thawing rate has been estimated to be over $1000°C$ min^{-1}. The capillary is opened at one end as shown in Fig. 1b. The small volume of cell suspension is aspirated with a sterile Pasteur pipette that has been drawn out very finely to a length of 4 cm. While aspirating the suspension, the pipette is slowly moved further into the capillary. A little experience is needed to avoid the uptake of air bubbles along with the cell suspension, which is then transferred to an opened culture tube containing fresh medium.

An alternative method for removing the cell suspension from a capillary is to use a 1 ml hypodermic syringe with a 0.65 mm needle. In this case the capillary is opened at both ends and the contents aspirated with a syringe previously flushed with oxygen-free gas. The needle is then inserted through the rubber closure of an unopened fresh culture tube. The tube is inverted and the suspension along with some medium is drawn into the syringe. Still in an inverted position, the contents of the syringe are ejected into the tube. This can be repeated once, but care should be taken not to expel gas bubbles into the tube if a plastic syringe is used.

IV. FREEZE-DRYING

Experience of freeze-drying methanogens is presently limited to only a few species. Studies performed on seven species of five genera have shown that members of the genera *Methanobacterium*, *Methanobrevibacter* and *Methanosarcina* in general survive the freeze-drying process better than members of *Methanococcus* and *Methanospirillum* (Hippe and Tilly, 1982). However, viable cultures could be recovered from them all after storage for 3–4 years. Examples of viability levels of some methanogenic strains after freeze-drying and storage at two different temperatures for up to 42 months are given in Table I. It can be seen that for some of the strains optimal

Table 1: Viability of some methanogenic bacteria after freeze-drying and storage.[a]

| Species | DSM no. | Freeze-drying[c] | | Storage at[c] | | |
		before	after	8°C 12 months	8°C 42 months	−70°C 42 months
Methanobacterium formicicum[b]	1312	2.9×10^7	4.8×10^3	—	1.3×10^1	2.3×10^3
Methanobacterium thermoautotrophicum	1053	1.8×10^8	8.0×10^7	1.8×10^4	5.5×10^5	3.6×10^5
Methanobrevibacter arboriphilus	1125	8.0×10^{10}	5.9×10^{10}	1.3×10^{10}	1.0×10^{10}	1.1×10^{10}
Methanobrevibacter ruminantium	1093	9.4×10^9	3.2×10^5	6.6×10^5	5.9×10^5	4.8×10^4
Methanobrevibacter smithii	861	2.0×10^9	1.3×10^8	2.9×10^7	1.1×10^8	1.1×10^9
Methanococcus vannielii	1224	1.0×10^7	6.4×10^3	5.0×10^3	0	2.4×10^3
Methanosarcina barkeri	804	6.9×10^7	1.2×10^6	1.0×10^5	5.3×10^3	5.0×10^6
Methanospirillum hungatei	864	4.4×10^6	8.0×10^3	8.0×10^3	3.1×10^3	2.0×10^4

[a] The suspending medium contained horse serum and 7.5% glucose mixture with 3 mg ml^{-1} ferrous sulphide.
[b] Suspending medium for *M.formicicum* contained no ferrous sulphide.
[c] The numbers indicate colony forming units per sample.

conditions of freeze-drying have not yet been achieved.

Suspending media, such as horse serum plus 7.5% glucose or 5% *m*-inositol, 10% skim milk plus 10% sucrose, or 12% sucrose alone, which are widely used in freeze-drying other bacteria, are also suitable for methanogens. Inclusion of small amounts of amorphous ferrous sulphide (usually 1 mg per ml) with the suspending medium gave good protection of the cells against oxygen and allowed distribution of the small volumes of suspension in air, rather than in inert gas.

The following procedure based on the double vial method has been developed for methanogens.

1. Preparation of Cell Suspension

Cultures are grown and cell suspensions are prepared as described in section III.A.1. The suspensions are kept under oxygen-free nitrogen gas until all vials have been filled.

2. Suspending Medium

The horse serum/sugar mixture is prepared by dissolving 7.5 g of glucose or 5 g of *m*-inositol in 100 ml of horse serum (Oxoid Ltd, SR 35, natural clot). The solution is sterilized through a Seitz filter-pad and 5 ml volumes are dispensed under oxygen-free nitrogen in sterile anaerobe test tubes. Before use, 0.3 ml of concentrated suspension (*c.* 20 mg ml^{-1}) of amorphous ferrous sulphide in water is added.

The ferrous sulphide is prepared according to Brock and O'Dea (1977). Sterile stock suspensions, after autoclaving in sealed tubes under nitrogen gas atmosphere, can be stored at room temperature for several months.

3. Filling Vials

Clean, cotton-wool plugged, labelled, sterile vials (11 × 45 mm; internal diameter 9 mm) are filled with 0.1 or 0.2 ml of cell suspension. Filling is carried out under air using a calibrated Pasteur pipette. An equal volume is used to inoculate a fresh culture tube for viability determination.

4. Pre-freezing

Suspensions are pre-frozen in a paraffin wax block before drying. The block (12 × 17 × 4 cm) is drilled with 24 holes 1.2 cm wide and

1.5 cm deep. The block, which is cooled overnight in a −20 or −70°C deep-freezer, is filled with up to 24 vials and replaced in the freezer for a further 15 min. Still in the wax block, the vials are transferred to the vacuum dryer. The cooled block prevents thawing of the cell suspension during transport and subsequent drying.

5. Freeze-drying

The entire paraffin wax block containing the vials is placed under the plexiglass bell of a freeze-drying machine. Primary drying is carried out until the vacuum has dropped to 0.05–0.01 torr and all free liquid has been removed. This takes about 4 h with 0.2 ml of cell suspension per vial. The drying process is then interrupted, the evacuated system is flooded with nitrogen gas and the vials are removed from the machine. The projecting parts of the cotton-wool plugs are cut off and each vial is placed in a second (outer) glass tube measuring 15 × 135 mm. The outer tubes contain a few pieces of self-indicating silica gel covered with a little cotton wool. The vials are covered with glass wool or stone wool slightly compressed to a layer 1–2 cm deep. The outer tubes are constricted above the inner vials and when cool are attached to the manifold of the freeze-drying machine for secondary drying overnight. At a vacuum of 0.001 torr, the tubes are heat-sealed at the level of the constriction. The ampoules are stored in the dark at room temperature or preferably in a deep-freeze cabinet.

6. Recovery and Viability Determination

To open the double-vial ampoule, the pointed part of the outer tube is heated in a Bunsen flame. One or two drops of water are added to crack the hot glass and the tip of the tube is knocked off. The inner vial is taken out and immediately gassed with the appropriate gas mixture (H_2/CO_2 or N_2/CO_2). About 0.5 ml of fresh medium from an opened gassed culture tube is added by means of a Pasteur pipette. The contents of the vial are agitated with the pipette to dissolve and resuspend the dried sample. After two minutes the contents are transferred to the culture tube. After sealing the inoculated tube, dilution series are prepared for viability determination as in section III.A.5.

11

Maintenance of Fungi

D. SMITH
Commonwealth Mycological Institute
Kew, UK

I. INTRODUCTION

Collections of fungi were originally kept by serial transfers from staled to fresh media. This method is still used for working

MAINTENANCE OF MICROORGANISMS
ISBN 0 12 410350 2

collections holding small numbers of cultures that are in constant use. However, this is a very time-consuming and labour-intensive exercise, especially when a large number of organisms is kept. Other problems associated with this method are that the cultures are often short lived and are liable to contamination both by other micro-organisms and by mites. Their morphology or physiology may change due to adaptation to growth on artificial media; for example, they may lose their ability to sporulate or reproduce sexually, or lose certain physiological properties. Again, plant pathogenic fungi grown on media for several generations often lose their pathogenicity and must be transferred to their host to check if this has occurred.

Over the years, several methods of preservation have been developed to eliminate problems encountered with serial transfers. One of the aims has been to extend the storage period between subculturing by allowing growth to continue at a reduced rate. This has been achieved by growing the fungus on a weak medium, by lowering the incubation temperature, or by growing under reduced oxygen tension. The latter can be accomplished by the addition of mineral oil to a culture growing on standard media at normal growth temperatures. Most of these techniques are still open to many of the problems encountered during normal serial transfers.

The methods that have proved most successful have been those that reduce metabolism to an extent that induces artificial dormancy. This is usually achieved by dehydration or freezing. Lyophilization, or freeze-drying as it is more commonly known, allows the dehydration of fungi to a level that halts metabolism. Dehydration is achieved by the sublimation of ice and is continued until a residual moisture content of between 1 and 2% is achieved. To halt metabolism by freezing, the organism must be taken to temperatures below $-130°C$ (Mazur, 1966). Liquid nitrogen at a temperature of $-196°C$ has proved very useful as a coolant and storage medium for organisms. Liquid air has also been used (Joshi *et al.*, 1974), though the risks involved are much greater because of potential explosion.

The choice of preservation method depends upon many factors. For example, the preservation requirements of a small teaching collection differ from those of a national depository. Again, taxonomic collections require methods that stabilize the morphology and physiology of the organism. Similarly, industrial collections place much emphasis on techniques that maintain genetic stability, especially for strains that have special features required for industrial processes, such as the production of anti-biotics and enzymes. Above all, the availability of facilities and funds can be a key factor in the choice of a particular preservation technique.

Many reviews cover the preservation techniques that are available for microorganisms (Fennell, 1960; Nei, 1964; Onions, 1971; Jong, 1978; Heckly, 1978), particularly the techniques of freeze-drying (Raper and Alexander, 1945; Last *et al.*, 1969; Rowe, 1971; Smith, 1983b) and liquid nitrogen storage (Hwang, 1960; Butter-field *et al.*, 1974). Genetic stability can be achieved by freeze-drying (Onions, 1971; Heckly, 1978) and liquid nitrogen storage (Prescott and Kernkamp, 1971) both of which prove to be excellent long-term storage techniques. However, for the successful use of any of the methods it is essential that the culture is in good condition. Optimum growth and sporulation will give better revivals after preservation; poor cultures are rarely improved by preservation.

II. MAINTENANCE BY SUBCULTURING

A. *On Agar*

Fungi can be grown on a wide range of agar media (Booth, 1971a). It is common practice to attempt to simulate the natural conditions of an isolate in order to obtain its fullest development. In many cases the relationship between the organism and others found in its natural environment can not be simulated in pure culture. However, a natural substrate can be provided and the use of sterile natural media or vegetable extracts may be helpful (Dade, 1960).

Some fungi have special requirements for successful storage and, wherever possible, these should be provided. Dermatophytes survive best on hair (Doory, 1968), some water moulds are best stored in water, often with the addition of plant material (Goldie-Smith, 1956) and other more sensitive water moulds may require aeration (Clark and Dick, 1974; Webster and Davey, 1976). It may be necessary to grow some fungi in mixed culture; *Dictyostelium*, for example, may require bacteria and the mycoparasite *Piptocephalis* can be provided with its microfungus host, *Cokeromyces* or *Penicillium*.

The culture vessel may be a Petri dish, a McCartney bottle, a plastic or glass universal bottle, or a test tube. In selecting a container various aspects of storage must be taken into consideration and storage space is perhaps the first selection criterion. The material from which the vessel is made is important if the isolate requires "black light" treatment (see section X), in which case plastic should be selected in preference to glass.

The type of inoculum can affect the quality of the fungal collection. It is thought that a mass spore transfer is by far the best technique and that mycelium transfer should only be used where no spores are available (Onions, 1971). There are occasions when this is not the case. Continual transfer of the spores of aging Mucorales can lead to the deterioration of the culture and transfer of spores, together with mycelium and substrate will avoid this problem. When mycelium transfer must be made it should be done from the growing edge of the colony. Pythiaceae are best subcultured by removal of the basal felt (Onions, 1971).

Single-spore cultures (Gordon, 1952) may be used with some unstable strains where with experience, the wild-type spore can be selected and transferred to fresh medium. This technique is also useful when separating an isolate from a more rapidly growing contaminant.

B. On Agar in the Refrigerator

The fungi are subcultured (see section II.A) and stored in a refrigerator at a temperature between 4 and 7°C.

C. On Agar under Oil

Healthy, mature cultures grown on slants of suitable agar in 1 oz universal bottles without cap liners are covered with twice auto-claved (121°C for 15 min) liquid paraffin (Medicinal paraffin, specific gravity 0.830–0.890) to a depth of 1 cm above the highest point of the fungal growth or agar slope. The bottles, with their caps loose, can be stored at room temperature or in an air-conditioned room maintained at 15–20°C.

Retrieval of a culture maintained in mineral oil is achieved by removing a small amount of the fungal colony on a mounted needle and draining away as much oil as possible on the neck of the culture bottle. The fungus is then streaked onto a suitable agar medium. More than one subculture may be necessary after retrieval as the growth rate can very often remain slow because of adhering oil.

D. On Agar Blocks in Water

The fungi to be stored are grown on suitable agar media in Petri dishes. Agar blocks, 6 mm^3, are cut from the growing edge of the fungal colony and transferred to sterile distilled water in McCartney bottles. The lids are tightly screwed down and the bottles stored at room temperature.

III. MAINTENANCE BY DRYING

A. On Anhydrous Silica Gel

1. Preparation of silica gel
Universal bottles are one-quarter filled with purified non-indicator silica gel, 6–22 mesh, and sterilized in an oven at 180°C for 2–3 h. The sterile gel is stored at 37°C, or in a desiccator, to avoid absorption of moisture; if this occurs the gel must be resterilized before use. The bottles are placed in trays of water to a depth above the level of the gel. The water is frozen by placing the trays in a deep freeze (-17°C to -24°C).

2. Suspending medium
A 5% solution of non-fat skimmed milk is prepared in distilled water and dispensed into universal containers in 5 ml amounts and autoclaved immediately at 115°C for 10 min. The skimmed milk solution is cooled to 4°C and stored at this temperature prior to use.

3. Spore suspension
Sporulating cultures of the fungi are placed in the refrigerator and cooled. The cooled skimmed milk is added to the cooled culture and a spore suspension is prepared by gently scraping the fungal colony with a sterile wire or glass rod.

4. Dispensing and drying the fungal suspension
The spore suspension is added to the cooled silica gel crystals in the trays of frozen water using a Pasteur pipette. Only three-quarters of the gel should be wetted to avoid over-saturation. The gel bottles are left in the ice bath for about 20 min until the ice around them has melted a little. The crystals are agitated to ensure thorough dispersion of the suspension. The bottles are kept at 25°C until the crystals readily separate when shaken, which is usually between one and two weeks.

5. Viability
The viability is checked by sprinkling a few crystals onto a suitable medium and assessing the amount of growth from each crystal. If growth is satisfactory the caps of the universal bottles are screwed down tightly and stored over indicator silica gel in an airtight container, such as plastic freezer boxes with good seals, at 4°C. The

indicator gel will require replacing once or twice a year. The "spent" gel can be dried out by heating at 180°C for 2 h or until the colour is fully restored.

B. In Soil/Sand

1. The soil
Garden loam, having a water content of approximately 20%, is put into glass universal bottles to between half and two-thirds capacity. This is sterilized by autoclaving at least twice at 121°C for 20 min, allowing the soil to cool between each autoclaving. Sand can be similarly sterilized and used instead of soil (Mehtra *et al.*, 1977).

2. The inoculum
A spore suspension is prepared by adding 5 ml of sterile distilled water to a culture and gently scraping the colony to release the spores. If the isolate does not sporulate in culture, a suspension of mycelium can be used.

3. Procedure
The suspension is dispersed in 1 ml amounts into the sterile soil or sand and a growth or drying-out period is allowed at room temperature. Two to three days is usually long enough for fast-growing isolates, but for most other isolates a period of 10–14 days allows the fungi to utilize the available moisture and for growth to slow down sufficiently for storage.

4. Storage and recovery
The soil/sand culture bottles are stored with loose caps in the refrigerator (4–7°C). The fungi can be revived by sprinkling a few grains of soil onto a suitable medium.

IV. MAINTENANCE BY FREEZE-DRYING

A. Method

1. Preparation of ampoules
The isolate number and date of processing are printed on the outside of (0.5 ml) glass ampoules (Anchor Glass Co. Ltd) using a reverse type printer (Rejafix Ltd). The ampoules are covered with lint caps and heat-sterilized at 180°C for 3 h. The sterilization will also dry and fix the ink.

2. Inoculum
A spore suspension of the isolate is prepared by adding 10 ml of a

10% w/v skimmed milk and 5% w/v inositol mixture (autoclaved at 115° for 10 min) to a mature sporulating culture grown on a suitable agar medium. The spores are suspended by gentle agitation and scraping of the fungal colony.

3. Inoculation of ampoules

Approximately 0.2 ml of the spore suspension is added aseptically to each sterile ampoule using a Pasteur pipette. The filled ampoules are loaded into the centrifuge racks of the EF6 freeze-dryer (Edwards High Vacuum).

4. Freeze-drying procedure

(1) The loaded centrifuge racks are lowered into the chamber of the EF6 freeze-dryer and the chamber closed. The centrifuge is switched on and the ampoules spun at 1425 rev/min. The chamber is evacuated and when a pressure lower than 0.5 torr is achieved the product will be frozen and the centrifuge is switched off. Drying is accelerated by heating the chamber walls and is carried out for a further 3 h.

(2) The chamber heaters are switched off and the chamber brought to atmospheric pressure. The ampoules are removed and plugged with sterile cotton-wool which is compressed to a depth of 1 cm and pushed down inside the ampoule until it is just above the slope of the dried suspension. The ampoules are then constricted using a Flame Master air/gas torch (Buck & Hickman) about 1 cm above the top of the cotton-wool plug.

(3) The constricted ampoules are attached to the manifold of an Edwards 30S2 secondary freeze-drying unit and evacuated over phosphorus pentoxide desiccant. The interval during which the freeze-dried material is exposed to oxygen and water vapour in the atmosphere must be kept to a minimum as over-exposure can cause deterioration (Rey, 1977). The drying period on the 30S2 is usually carried out overnight for 17 h leaving a residual moisture content of between 1 and 2%.

(4) The ampoules are sealed under vacuum at the point of constriction using a crossfire burner (Wesley Coe (Wingent) Ltd; Adelphi Manufacturing). The ampoules are stored in an air-conditioned room maintained between 15 and 20°C. After 3–4 days' storage an ampoule is opened and the fungus reconstituted by the addition of 3–4 drops of sterile distilled water from a Pasteur pipette and allowing 15–20min for absorption of moisture by the spores. The contents of the ampoule are streaked onto a suitable agar medium and successful retrieval is determined by resumption of normal

growth and sporulation.

V. MAINTENANCE BY FREEZING

A. *On Agar Slopes at* $-20^\circ C$

The cultures are prepared as for normal transfer (see section II.A) and when mature they are placed in a deep-freeze at $-20°C$. A small portion of the frozen colony is removed and placed on a suitable agar medium and allowed to thaw at room temperature. The frozen stock culture is replaced in the freezer as quickly as possible to avoid thawing.

B. *In Liquid Nitrogen*

1. Preparation of ampoules
The culture number is written, using a permanent felt tip pen (Scotts Office Equipment Ltd), or printed, using a reverse type printer (Rejafix Ltd), on to 1.0 ml borosilicate glass ampoules (Anchor Glass Co. Ltd). The ampoules are covered with individual tin-foil caps and sterilized in an oven at 180°C for 3 h.

2. Culture preparation
The fungi are grown on slopes of suitable agar in 1 oz universal containers (Sterilin Instruments). After growth is well established the cultures are transferred to a refrigerator at 4–7°C to allow continuing growth and a degree of cold hardening. Not all fungi will grow at these temperatures and for such isolates the cold-hardening stage may be entirely omitted or may be reduced to a short storage period prior to freezing. This stage is isolate dependent and can be determined by pre-growth tests before storage in liquid nitrogen.

3. Preparation of inoculum
10 ml of 10% v/v glycerol is dispensed into glass universal bottles and autoclaved at 121°C for 15 min, cooled, and stored at room temperature prior to use. The glycerol is added to the culture and the mycelium or spores suspended by gentle agitation and scraping of the fungal colony.

4. Inoculation, sealing, and testing of ampoules
The fungal inoculum is added in 0.5 ml amounts to the sterile borosilicate glass ampoules and heat-sealed using a flame-master air/gas torch. The seals are tested by immersing the ampoules in an erythrosin B dye bath at 4–7°C, which also pre-cools the fungal

suspensions.

5. Freezing protocol

The ampoules containing the fungal suspension are frozen at approximately $1°C\ min^{-1}$ by suspension in the vapour phase of a liquid nitrogen refrigerator at $-35°C$ for 40–45 min. This is followed by plunging the ampoules into liquid nitrogen which cools them to $-196°C$.

6. Storage and viability

The ampoules are stored either clipped to aluminium canes in boxes in a 250 l refrigerator (Union Carbide) or in the drawer rack system of a wide neck 320 l (Union Carbide) refrigerator. To check the viability of cultures the ampoules are removed from the liquid nitrogen refrigerator and thawed by immersion in a water bath at 37°C. When the suspension has thawed it is streaked onto a suitable agar medium, incubated and examined for growth.

7. Special treatments

Some fungi are damaged by excessive manipulation. To avoid this they are placed in the ampoule on small slivers of agar. Alternatively, they may be grown in the ampoules on a small amount of suitable agar medium. If the culture does not survive using glycerol as the cryoprotectant, 10% dimethyl sulphoxide (DMSO) or a mixture of 5% DMSO and 8% glucose (Smith, 1983a) may be used.

VI. STORAGE AND SURVIVAL IN SUBCULTURE

A. *On Agar at Growth Temperatures*

The storage conditions of growing cultures are often dependent on the physiological requirements of each fungus. In addition to providing a suitable growth medium, temperature, light and humidity must also be taken into account.

The growth of cultures at room temperature is perhaps the most convenient method of maintenance. Keeping culture vessels in racks and storing them on shelves between transfers is suitable for the majority of microfungi, which normally grow well at temperatures of 20–22°C. However, some fungi are more sensitive to temperature and it is necessary to establish the range in which an isolate will grow before selecting a suitable storage temperature. Chytrids and other water moulds grow better at temperatures lower than 20°C whereas some thermophiles require temperatures between 30–50°C for growth.

The light requirement of a fungus for spore formation will influence the choice of storage position. Some fungi, such as *Pyronema domesticum*, require daylight to induce production of fruiting bodies and sporulation. Others, such as some species of *Coprinus*, respond better to growth in the dark and produce sporophores. Many Dematiaceous Hyphomycetes, Coelomycetes, and Ascomycetes require the stimulation of near ultraviolet or black light before they will produce spores (Leach, 1971).

Most fungi grow well at a fairly high humidity (Onions, 1971) and limiting water availability will certainly hinder their growth. Normally, within culture vessels the moisture in the atmosphere above the culture provides an adequate climate.

A collection of growing fungi must be checked at regular intervals to ensure that they have not become desiccated. The rate at which the cultures lose moisture will depend upon the conditions under which they are stored. High temperatures and low humidities will inevitably mean that transfers should be more frequent. The Centraalbureau voor Schimmelcultures (CBS), Netherlands, recommends temperatures of 16–17°C and a relative humidity of 60–70% for storage (von Arx and Schipper, 1978). Keeping cultures in high humidities will allow condensation and growth on the outside of culture vessels and will lead to cross contamination.

Transfers to fresh media are normally made every 3–6 months. Some cultures, such as the water moulds and human and animal pathogens (Onions, 1971), require more frequent transfer and collections kept in the tropics or low humidity climates require even more frequent attention (Fennell, 1960). The genera *Allomyces*, *Achlya*, *Isoachlya*, *Phytophthora*, *Pythium* and *Saprolegnia*, and the basidiomycete genera *Boletus*, *Coprinus*, *Corticium*, *Cortinarius* and *Mycena* require subculturing either once a month or every 2–3 months (von Arx and Schipper, 1978).

B. *On Agar in the Refrigerator*

The use of cold storage to slow the rate of metabolism, which increases the period between transfers to fresh media, has proved very successful. Normal domestic or laboratory refrigerators, which give temperature ranges of 4–7°C, may be used. The cultures are stored in racks or boxes on the shelves to give easy access. Overpacking of refrigerators that are in regular use can cause problems due to build-up of condensation and the probability of cross-contamination (Dade, 1960). A yearly transfer is usually adopted for these cultures, though species which fail to survive this storage period must be transferred more frequently. Most moulds and common fungi, such as the penicillia, aspergilli and Mucorales, survive well at 5–8°C but some, such as *Piptocephalis*, are sensitive to cold (Onions, 1971). A decline in the antibiotic titre of *Penicillium chrysogenum* (MacDonald, 1972; Wellman, 1971) has been observed with storage at 4°C though the growth and sporulation of this species seems to be unimpaired at this temperature.

At the Commonwealth Mycological Institute (CMI) all cultures are kept at 4–7°C, pending the results of storage by other means. Storage periods are normally between 4 and 6 months, but many cultures have survived for 1–2 years when left in the refrigerator. Although specimens have been shown to dry out after 9 months, this depends on the medium on which the fungus is growing. In general, the Zygomycotina, Ascomycotina, Basidiomycotina and Deuteromycotina successfully survive storage at 4–7°C for periods up to 1 year, whereas the Mastigomycotina survive less well.

C. On Agar under Oil

Using the method reported by Buell and Weston (1947) the entire collection at CMI was stored under oil following trial experiments in 1950 (Table I). Since 1956 the number of isolates stored by this method has increased to over 10,000. Although many have survived storage for over 25 years, a number have deteriorated or died. Forty genera require subculturing every two years because they have shown deterioration during storage (Table II). Some fungi, such as strains of *Fusarium*, some Saprolegniaceae and other water moulds, are not stable or do not survive long storage periods under oil and so require subculturing at 3–6 month intervals. However, it has been reported by others (Reischer, 1949) that Saprolegniaceae and other water moulds survive 12–30 months. As with most storage techniques survival seems to vary for individual strains within a species. The frequency of transfer is also affected by the poor condition of the isolate or failure to sporulate in culture.

Although storage under mineral oil has the advantage of being inexpensive and simple it has the major disadvantage of allowing growth under what may be adverse conditions. The slower growth rate induced may allow naturally occurring mutants to become dominant over the wild type. The occurrence of sectoring as an expression of mutation is increased after storage under a layer of mineral oil. Constant supervision is required by a specialist to ensure that the original strain, rather than the mutant, is transferred.

Table I: Survival of some cultures from the 1950 CMIoil collection.

	IMI number	Survival (years)
Aspergillus avenaceus	16140	32
A. citrisporus	25285	32*
A. medius	29188 ii	20
A. restrictus	39044	24
Botryosphaeria obtusa	38560	32
B. ribis	36476	32
Ceratocystis paradoxa	37270	32
C. paradoxa	39075	32
C. radicicola	36479	32*
Chlamydomyces palmarum	39639	32*
Corticium praticola	34886	32
C. rolfsii	33912	32
C. rolfsii	37955	25*
Drechslera halodes var.		
tritici	22971	32
Eleutherascus terrestris	25845	32*
Helicodendron triglitziense	38968	32*
Helminthosporium portulacae	37710	32*
Humicola sp.	38777	32*
Mortierella alpina	38598	12 D
Mucor circinelloides	39478	23
Nectria pityrodes var.		
saccharina	37228 a	32*
Penicillium adametzii	39751	25*
P. aurantioviolaceum	39740	32
P. cyaneum	39744	25
P. ehrlichii	39737	32*
P. fellutanum	39734	32
P. fuscum	39747	32
P. javanicum	39733	32*
P. lapidosum	39743	32
P. levitum	39735	32
P. lividum	39736	32
P. oxalicum	39750	32
P. sclerotiorum	39742	32
P. shearii	39739	32
P. striatum	39741	32
P. trzebinskii	39749	32
P. turbatum	39738	32
P. waksmanii	39746	25 (sectoring)
Petriella sordida	38601	32
Phytophthora citricola	21173	32

Table 1—continued

	IMI number	Survival (years)
P. nicotianae	22176	32
Podospora fimbriata	38111 i	10 D
P. fimbriata	38111 1ii	32
P. fimbriata	38111 2i	32
P. fimbriata	38111 2ii	32
Rhizoctonia oryzae-sativae	31287	32
R. solani	20697	32
Sclerotium coffeicola	39753	32*
Scopulariopsis sp.	16401	32*
Scopulariopsis sp.	16404	32*
Sporendonema casei	37084	17 D
Stephanosporium cereale	38105	32
Thielaviopsis basicola	36482	12 D
Torula herbarum	31291	32
T. ligniperda	36123	32
Ustilago scitaminea	36859	32
Verticillium theobromae	31432 a	32
Volutella ciliata	38780	32

* Showed inconsistency on retrieval and required several isolations to achieve good growth.
D—Died between number of years' survival indicated and the next viability test.

D. On Agar Blocks in Water

This simple method of storage has been used for the maintenance of fungi for many years. Fungi pathogenic to humans (Castellani, 1939; Castellani, 1967), plant pathogenic fungi (Figueiredo, 1967; Figueiredo and Pimentel, 1975; Boeswinkel, 1976), Oomycetes (Clark and Dick, 1974), Entomophthorales, Pyrenomycetes, Hymenomycetes, Gasteromycetes, Hyphomycetes (Ellis, 1979) and ectomycorrhizal fungi (Marx and Daniel, 1976) have all survived using this technique.

At CMI the method has been used successfully with the genera *Phytophthora* and *Pythium*, both of which have survived 2 years' storage without loss in viability.

VII. STORAGE AND SURVIVAL OF DRIED CULTURES

A. On Anhydrous Silica Gel

Many fungi have been successfully stored by this technique (Onions,

Table II: Genera and number of isolates stored under mineral oil in the CMI Culture Collection that require regular transfer every two years because of deterioration during storage.

Genus	No. isolates	Genus	No. isolates
Acremoniella	1	Helicosporina	1
Alternaria	4	Heliscus	1
Arthobotrys	14	Hypomyces	3
Basidiobolus	7	Khuskia	4
Beltrania	4	Lophiostroma	2
Beltraniella	1	Melanospora	2
Calonectria	3	Mesobotrys	1
Cercospora	16	Monacrosporium	8
Chaetosphaeria	5	Monascus	4
Chalaropsis	1	Mortierella	58
Chrysosporium	1	Mycovellosiella	5
Claviceps	3	Nectria	5
Colletotrichum	81	Nodulisporium	41
Cylindrocarpon	3	Penicillifer	2
Cystospora	1	Periconia	24
Echinosporangium	1	Pseudocercospora	1
Endophragmia	1	Pyrenochaeta	5
Georgefischeria	1	Pyrenophora	24
Gloeosporium	4	Spermospora	1
Glomerella	8	Stigmina	1

1977; Smith and Onions, 1983) but it is not satisfactory for mycelial cultures or for species belonging to the Mastigomycotina. It has proved to be a good technique for maintaining cultures in a stable condition since many strains of genetic importance have been preserved without change (Ogata, 1962; Perkins, 1962; Barratt *et al.*, 1965). The pathogenicity of *Helminthosporium* (*Drechslera*) *maydis* was maintained for 1 year using this method (Sleesman *et al.*, 1974).

The storage of 426 isolates was examined at CMI and the results, broken down into major taxonomic groups according to Ainsworth (1971) and the Index of Fungi (Anon, 1971–1981), are shown in Table III. Although 308 (72%) survived for periods up to 11 years, 118 failed to survive (Smith and Onions, 1983). It appeared that survival was isolate specific.

The 118 failures have survived other storage techniques and a list of these isolates is given with their survival period in Table IV. The limiting periods for survival are not yet known because storage continues.

Table III: Viability of all the isolates processed and stored on silica gel at CMI from 1971 to 1982.

	Number of Genera Viable	Number of Species Viable	Number of Isolates Tested	Viable	% viability
MYXOMYCOTA					
Acrasiomycetes	0	0	1	0	0
EUMYCOTA					
MASTIGOMYCOTINA					
Chytridiomycetes	0	0	5	0	0
Oomycetes	0	0	5	0	0
ZYGOMYCOTINA					
Zygomycetes	9	17	34	20	59
ASCOMYCOTINA					
Hemiascomycetes	3	4	5	5	100
Plectomycetes	5	9	13	11	85
Pyrenomycetes	17	23	54	35	63
Discomycetes	2	2	6	2	33
BASIDIOMYCOTINA					
Hymenomycetes	8	10	19	10	52
Gasteromycetes	0	0	1	0	0
DEUTEROMYCOTINA					
Hyphomycetes	38	136	250	202	81
Coelomycetes	7	18	28	21	75
ACTINOMYCETES	1	1	3	1	33
LICHENS	1	1	2	1	50

Viability of all isolates tested from 1971–82

Tested	Failed to survive during storage	Viable	% viability
426	118	308	72

B. In Soil/Sand

Fusarium, a genus of fungi renowned for its instability in culture, has been kept in a stable condition by storage in soil (Gordon, 1952; Booth, 1971b). *Septoria* species have been kept by this method without loss of sporulation or pathogenicity (Shearer *et al.*, 1974) and storage of *Pseudocercosporella herpotrichoides* has been very successful (Reinecke and Fokkema, 1979.)

At CMI the technique has been used successfully for 20 years to

Table IV: Survival, using various methods, of the 118 fungi that failed to survive storage in silica gel from 1971–82.

| | IMI Number | Survival Periods in Years | | | |
		SG	Oil	FD	LN
Allomyces arbuscula	129543	—	14	—	7
A. arbuscula	152201	5	7D	0	0
*A. javanicus**	144364	—	4D	0	0
Armillaria mellea	61755	—	27	0	0
Arnium arizonense	169785	—	8	8	8
Arthrobotrys oligospora	102121	<3	16	6	0
Ascochyta fabae	135517	—	7P	12P	0
Ascotricha lusitanica	147693	—	11	11	11
Aspergillus candidus	130667	<3	14	8	0
A. citrisporus	25285	4	11P	13	0
A. ochraceus	16247iii	2	11	11P	0
A restrictus	127782	<3	14	14	0
A. ustus	100391	3	18	8	0
A. wentii	162039	4	11P	1	0
Basidiobolus meristosporus	108476	<3	12	5	0
Beltrania sp.	223748	—	4	2	0
Botryodiplodia theobromae	125847	4	9	8	0
Calonectria quinquiseptata	136139	<3	11	—	11
Candida lipolytica	93743	—	20	15	0
Ceratocystis ulmi	147188	2	9P	1	0
C. ulmi	173135	—	9	6	0
C. ulmi	173136	3	9	9	0
*Cercospora beticola**	77043	—	4P	0	0
C. sesami	111779	—	12P	12ST	12
*Chytridium olla**	86666	—	7D	0	—
Cochliobolus sativus	166172	—	9	9	9
C. sativus	166173	—	9	9	9
Coemansia formosensis	170166	—	9	9	8
*C. mojavensis**	140079	1	0	0	0
C. pectinata	142377	2	10P	9	10
Colletotrichum gossypii	82269	2	5ST	12	12
C. trichellum	82378	<4	15	10	0
C. truncatum	86431	1	21	5	0
Conidiobolus coronatus	68174	2	13	—	0
Conidiobolus heterosporus	102043	—	18	0	0
C. lobatus	138635	—	12	—	0
C. mycophagus	113701	<3	6	—	9
*Coprinus luteocephalus**	161421	—	0	9	0
*C. viarum**	161423	—	0	9	0
Corticium rolfsii	77445a	—	22	5	0
Culicinomyces clavosporus	177011	<6	6	6D	0
Cryptospora suffusa	173497	<3	9	9	9

Table IV—continued

	IMI Number	Survival Periods in Years			
		SG	Oil	FD	LN
Dictyostelium discoideum	69094ii	—	27	15	11
Exobasidiellum culmigenum	136517	—	13	7D	0
Fusarium culmorum	175485	<7	9	9	0
F. solani	76761	1	21D	15	0
F. solani	172507	—	0	7	0
Gäeumannomyces graminis	160145	—	11	2	2
Gelasinospora sp.	47702	—	24	13	0
Gnomonia fructicola	164147a	—	9	9	0
Helicodendron tubulosum var.					
phialosporum	92743	<3	12	6	10
Helicosporina veronae	114458	—	16	4	8
Heliscus submersus	82609	<3	16	9	9
Heterocephalum aurantiacum	131684	8	12P	12	14
Hypoxylon mediterraneum	75991	<3	22	0	0
H. nummularium	146051	0.2	6	—	0
Leptosphaeria doliolum subsp.					
pinguicola	199777	<4	6	0	5
Martensiomyces pterosporus	60573	—	26	2	2
Metarhizium anisopliae	98375	<1	13P	10	0
Micromonospora vulgaris	126892	5	15P	11	0
Monacrosporium salinum	109555	—	17	5	5
Mortierella bainieri	167609	—	9	9	0
Mycosphaerella deightonii	119431	—	7	—	0
Mycovellosiella ferruginea	124973	—	15	9	0
Neurospora crassa	19419	5	12	6	0
N. crassa	68614ii	—	12	12	12
N. crassa	147001	2	5D	0	0
Nomuraea atypicola	186963	<1	8	7	0
Penicillium brevicompactum	17456	<3	25	15	0
P. canescens	149218	—	11	11	0
P. corylophilum	101082	2	19	11P	0
P. corymbiferum	68414	<3	22	12	0
P. cyclopium	19759	<3	19	15	0
P. digitatum	91956	—	14	14	0
P. digitatum	92217	<3	12	14	0
P. expansum	191205	2P	7	7	0
P. helicum	197479	2P	7	6	0
P. idahoense	148393	—	12	12	12
P. janthinellum	108033	<4	18	15	0
P. lavendulum	40570	1	27	15	0
P. luteum	112513	—	5D	11	0
Phycomyces blakesleeanus	118496	5	12P	8	0

Table IV—continued

	IMI Number	Survival Periods in Years			
		SG	Oil	FD	LN
P. blakesleeanus	118497	11	11	12	0
Phoma epicoccina	164070	—	9	9	9
Phytophthora cactorum	21168	—	32	—	10
Piptocephalis xenophila	156650	4	10P	5P	11
Pleospora infectoria	173200	3	9	9	9
Pyrenopeziza brassicae	204290	—	0	3	0
Pyrenophora graminea	129760	<1	15ST	10D	0
Pyronema domesticum	57472	—	27	15	2
Pythium debaryanum	48558	—	28	0	0
P. flevoense	176046	—	8	0	7
P. middletonii	42098	—	31	0	10
Rhizoctonia carotae	162910	—	10	0	0
Rhizophydium sphaerotheca	143633	—	12	—	—
Rhizopus rhizopodiformis	158738	—	11	11	0
Rhodotorula rubra	38784	4	12	5	0
Ryparobius polysporus	75299	—	13	10	0
Saprolegnia ferax	146489	—	11	—	3
Sclerotinia sclerotiorum	147201	—	12	0	12
Sclerotium delphinii	159926	—	10	4	9
Seiridium sp.	151978	8	11	10P	0
Serpula lacrimans	152233	—	11	0	0
Sphaerobolus stellatus	155101	—	10	0	10
Sporobolomyces roseus	43529	—	31	11	0
Stachybotryna columnare	158980	—	10	0	0
Stilbum macrosporum	163252	—	9	9	0
Streptomyces griseus	50967	—	29	15	0
Streptomyces lisandri	137178	—	12	10	0
Syzygites megalocarpus	231978	<2	0	2	4
Thielavia fimeti	116692	—	16	10	0
T. terricola	153731	—	1D	11	11
Trichoderma viride	57421	5	25	4	0
Trichothecium roseum	129425	—	14	14	0
Tritirachium roseum	169856	<3	10	10	0
Zalerion maritima	81620	<3	14	—	10
Zopfiella leucotricha	153733	—	1	10	10

SG—silica gel; FD—freeze-drying; LN—liquid nitrogen storage.
D—dead at time of examination.
P—poor revival at time of examination.
ST—sterile at time of revival.
* These isolates were discarded from the collection after the initial processing as they were atypical.

store 764 isolates of *Fusarium* and other genera (Booth and Butterfill, unpubl.) (Table V). Not all of the isolates have been tested recently, but of 54 examined after 10–20 years' storage 48 gave good growth and six had died. Of the isolates that have died in storage all had survived for more than 10 years. It is felt that the considerable time lag before the onset of dormancy due to dryness might be sufficient for mutant vegetative strains to overgrow the wild type. An examination of a collection developed by Gordon (1952) showed that 76% of *Fusarium equiseti* isolates, 75% of *F. semitectum* and 50% of *F. acuminatum* had been outgrown by mutant strains (Booth, 1971b).

Table V: Isolates in the CMI collection that have been stored in soil.

Genus	Number of Species	Number of Isolates
Calonectria	6	8
Cylindrocarpon	12	32
Cylindrocladium	5	9
Fusarium	55	652
Giberella	4	6
Melanospora	2	9
Nectria	17	47
Thielavia	1	1

VIII. STORAGE AND SURVIVAL OF FREEZE-DRIED CULTURES

Although many fungi have been preserved by freeze-drying (Smith, 1983b) it is a technique that seems to be suitable only for sporulating isolates. Mycorrhizal fungi generally give poor results (Jackson *et al.*, 1973; Crush and Pattison, 1975), although with a modified technique cultures that are normally difficult to preserve can survive. Pre-freezing to −45°C of plant tissue infected with *Pythium acanthicum* and *P. irregulare*, followed by drying under vacuum at room temperature, facilitated the preservation of the pathogens (Staffeldt and Sharp, 1954). *Puccinia* urediniospores, prefrozen to between −45°C and −50°C and dried under vacuum at −10°C, survived freeze-drying (Sharp and Smith, 1952).

At CMI 8166 isolates have been freeze-dried, of which 7489 have survived. A list of genera and numbers of isolates that have failed to survive centrifugal freeze-drying is given in Table VI. As with most preservation techniques success with freeze-drying varies between isolates of the same species, although preservation is more likely to be successful if good healthy cultures are used. A list of genera that have survived 14 years' storage (Table VII) and those that have survived the technique without suspending medium on agar blocks (Table VIII) indicate the success of the technique.

Table VI: Genera that have failed to survive centrifugal freeze-drying.

Genus	Genus	Genus
Achlya	Herpotrichia	Pythium
Allomyces	Kretzschmaria	Quarternaria
Areolospora	Lacellinopsis	Saprolegnia
Armillaria	Lasiobolidium	Searchomyces
Arthrocladium	Lentinus	Selenosporella
Ascocalvatia	Lenzites	Selinia
Balansia	Leptoporus	Sigmoidea
Battarraea	Lomachashaka	Sphaerobolus
Biscogniauxia	Marasmius	Sphaerostilbe
Blastocladiella	Melanconis	Spondylocladiopsis
Calospora	Monotosporella	Stereum
Camposporium	Nummularia	Sympodiella
Chytridium	Panus	Syzygites
Cladobotryum	Penicillifer	Tetracladium
Coriolus	Phaeoisariopsis	Tetranacrium
Dactuliophora	Phyllosticta	Umbelopsis
Eleutherascus	Physarum	Urohendersonia
Entomophthora	Phytophthora	Ustilaginoidea
Eremomyces	Piedraia	Ustulina
Fomes	Platysomum	Volvariella
Ganoderma	Puccinia	

Table VII: Genera that have survived 14 years since freeze-drying at CMI.

Genus	Genus	Genus
Absidia	Elsinöe	Phialocephala
Acremonium	Eremothecium	Phialophora
Actinomucor	Fusarium	Phoma
Alternaria	Gelasinospora	Phomopsis
Amorphotheca	Geomyces	Phycomyces
Arthroderma	Geosmithia	Pirella
Ascotricha	Geotrichum	Pithomyces
Aspergillus	Gilbertella	Pleurophragmium
Aureobasidium	Gliocladium	Rhinotrichum
Bispora	Gliomastix	Rhizopus
Botryotrichum	Glomerella	Saccharomyces
Byssochlamys	Hirsutella	Scopulariopsis
Cephaliophora	Humicola	Sesquicillium
Cephalosporium	Leptographium	Setosphaeria
Cercospora	Loramyces	Spegazzinia
Chaetodiplodia	Mammaria	Sporobolomyces
Chaetomidium	Mariannaea	Sporophora
Chaetomium	Memnoniella	Sporothrix
Chaetopsina	Microascus	Stachybotrys
Chalara	Monodictys	Staphylotrichum
Chloridium	Mucor	Stemphylium
Circinella	Mycosphaerella	Stilbella
Cladosporium	Myrothecium	Streptomyces
Coniella	Nectria	Sydowia
Coniothyrium	Neocosmospora	Syncephalastrum
Cordyceps	Neodeightonia	Thamnostylum
Coryne	Neurospora	Thermoascus
Cunninghamella	Oidiodendron	Thielavia
Curvularia	Ophiobolus	Trichoderma
Cylindrocarpon	Paecilomyces	Trichophaea
Didymella	Penicillium	Ulocladium
Didymosphaeria	Pestalotiopsis	Valsa
Doratomyces	Petriellidium	Venturia
Drechslera	Phaeotrichoconis	Volutella
Eladia	Phialomyces	Zygorhynchus

More detailed information is available from CMI (Freeze drying at CMI, D. Smith, 1982).

IX. STORAGE AND SURVIVAL OF FROZEN CULTURES

A. *On Agar Slopes at −20°C*

Many cultures frozen and stored at −20°C can be expected to survive 4–5 years if they are not allowed to thaw. Some fungi are sensitive to this form of storage and die when frozen to this temperature. *Martensiomyces*, some Oomycetes, and water moulds fall into this group. The CMI taxonomic collection of *Aspergillus*, *Penicillium*, and related genera are kept by this means and have survived for 5 years.

Cold storage of fungi has been used for medically important fungi (Carmichael, 1956, 1962) and proves to be even more successful if the stock cultures are not allowed to thaw during retrieval (Kramer

Table VIII: The survival periods of isolates freeze-dried without suspending medium on agar blocks.

Isolate	IMI No.	Survival Period (yrs.)
Acremonium sp.	55286	14
Ascocoryne sarcoides	68130	14
Aspergillus amstelodamii	71295	8
Aspergillus candidus	73074	14
Aspergillus carneus	73777	14
Aspergillus nidulans var. *echinulatus*	61454ii	14
Aspergillus niger	75353ii	14
Aspergillus quadrilineatus	72733	14
Aspergillus ostianus	93445	14
Chaetomium abuense	114513	14
Curvularia trifolii f.sp. *gladioli*	75377	13
Cylindrocarpon congoense	69504	14
Embellisia chlamydospora	67737	14
Fusarium graminearum	69695	14
Nectria gliocladioides	71095	14
Paecilomyces dactylethromorphus	65752	14
Penicillium cyclopium var. *echinulatum*	68236	14
Penicillium nigrans	96660	14
Penicillium paraherquei	68220	14
Penicillium raperi	71625	13+
Penicillium roquefortii	129207	14
Penicillium spinuloranigenum	68617	14
Penicillium steckii	72029	14
Pestalotiopsis gracilis	69749	14
Phaeotrichoconis crotalariae	69755	14
Phialomyces macrosporus	110130	14
Phomopsis oncostoma	68344	14
Pycnoporus sanguineus	75002	9
Sagenomella griseobiridis	113160	13+
Scopulariopsis carbonaria	86941	14
Sporidesmium flexum	246524	1

and Mix, 1957). Freezing and thawing microbes has many associated problems (Calcott, 1978) and it is found that some fungi are susceptible to freezing damage without cryoprotectants.

B. *In Liquid Nitrogen*

Freezing and storing fungi in liquid nitrogen at CMI has proved very successful. Of the 3286 preserved in this way, 2900 have survived up to 14 years. The Mastigomycotina proved to be most difficult to preserve by this method, although the isolates that grew well in culture survived well (Table IV). This group of fungi seem to suffer from mechanical damage incurred during the preparation of suspensions. If precautions are taken not to cause damage of this kind the fungi survive well (Smith, 1982). Pre-growth of cultures in the refrigerator at 4–7°C can improve the viability of some fungi that are normally difficult to freeze.

Similar methods of storage in liquid nitrogen have proved successful for a wide range of fungi (Hwang, 1966, 1968; Hwang *et al.*, 1976; Butterfield *et al.*, 1974; Alexander *et al.*, 1980). Fungi that are difficult or impossible to grow in culture can be kept alive for long periods in liquid nitrogen. Examples are rust and smut spores (Loegering, 1965; Kilpatrick *et al.*, 1971) and *Sclerospora* species (Gale *et al.*, 1975; Long *et al.*, 1978; Smith, 1982).

There are many storage containers other than glass ampoules that can be used successfully in liquid nitrogen vapour. When polypropylene ampoules were used, better revival of *Basidiobolus, Rhizophydium, Phytophthora* and *Serpula* was achieved than with glass ampoules (Butterfield *et al.*, 1978). Although plastic ampoules with screw caps may leak during storage they are a safer alternative to glass because they do not explode on expansion of liquid nitrogen (Simione *et al.*, 1977). The use of plastic drinking straws that can be autoclaved are a space saving alternative (Dietz, 1975; Elliott, 1976) although storage conditions must not allow contact with liquid nitrogen, which may penetrate and cause contamination or splitting of straws on warming. Polyester film packets containing fungal material can also be used (Tuite, 1968).

X. BLACK LIGHT: ITS USE TO INDUCE SPORULATION IN FUNGAL CULTURES

Successful fungal preservation sometimes depends on the presence of spores. Light is very important in this respect and short wavelengths have been used to induce spore production (Leach, 1962, 1971). Near ultraviolet light or black light of wavelength 3100 to 4000 Å may affect pigmentation, the gross morphology of the colony, or even spore morphology, though the effects are not sufficient to interfere with identification.

The light-benches at CMI have three 1.20 m fluorescent lamp holders 13 cm apart. A black-light tube (Philips TL 40 W/08) is held in the centre holder and a cool-white tube (Philips MCFE 40 W/33) is placed on either side. They are controlled by a time switch which is set at a 12 h on/off cycle. The fungi in polystyrene disposable Petri dishes are supported on a shelf 32 cm below the light source. The fungi are inoculated onto the plate and allowed to grow for 2–4 days before placing under the light. The edges of the dishes are sealed with transparent adhesive to prevent rapid drying (Booth, 1971b).

The growth medium can have an effect on stimulation and weak media should normally be used to encourage sporulation (Leach, 1962). When growing on glucose casein hydrolysate medium and stimulated by near ultraviolet light, *Diaporthe phaseolorum* v. *batatis* produces many perithecia and ascospores. However, when grown on malt or potato glucose media the number of perithecia is much less (Timnick *et al.*, 1951).

XI. MITE PREVENTION

Mites, commonly of the genera *Tyroglyphus* and *Tarsonemus*, can cause damage to fungal cultures in two ways. First, they can eat the culture and completely destroy it. Secondly, they may carry fungal spores and bacteria on their bodies and, as they move from one culture to another, cross-contamination may occur. Methods of control used by different workers are varied and a combination of precautions seems to be most effective. Hygiene, fumigation, mechanical, and chemical barriers and protected storage are the normal methods of prevention (Anon, 1982).

A. *Hygiene*

Hygiene coupled with a quarantine procedure is perhaps the best protection. All work surfaces should be kept clean and cultures should be protected from aerial infection by storage in cabinets. Benches should be washed down with acaricides. Kelthane (Murphy Chemical Co.), tedion V-18 (Mi-Dox Ltd), chlorocide (Boots Pure Drug Co. Ltd) and paradichlorobenzene (BDH

Chemicals Ltd) are suitable. A "dirty" room should be available to which all incoming cultures are directed to ensure that they are mite-free. When mites are found the infested cultures should, ideally, be sterilized and fresh isolates sought. If this is impossible the culture should be covered with a quantity of liquid paraffin and subcultured at a later date.

B. Fumigation

Unpurified tractor vapourizing oil was formerly used at CMI and, although not an acaricide, it deters mites. Paradichlorobenzene is effective, although it can produce abnormal growth in fungi, and camphor is another fumigant that can be used successfully (Smith, 1967). Fumigants are placed in the storage cabinets either as a short treatment or for permanent storage, if the chemicals used do not affect the fungi.

C. Mechanical and Chemical Barriers

Surrounding cultures with oil, water or vaseline can prevent infection from crawling mites, although those carried on clothing, by insects or by air currents can still cause infestation.

Sealing the necks of tubes or bottles with sterile cigarette paper attached with copper sulphate gelatine glue (20 g gelatine, 2 g copper sulphate in 100 ml distilled water) can prevent mite infection and still allow air to pass thus not impeding fungal growth (Snyder and Hansen, 1946). Smith (1971) recommends the use of disposable plastic bottles with plastic caps that can be screwed down tightly to prevent the entry of mites.

D. Cold Storage

Although cold storage at 4–8°C reduces the spread of mites, they soon become active on removal from the refrigerator. Deep-freezing below −18°C will kill most mite eggs.

12

Maintenance of Yeasts

B. E. KIRSOP
National Collection of Yeast Cultures
Food Research Institute
Norwich, UK

I. INTRODUCTION

In general, yeast cells are considered to be robust, tolerant of unfavourable conditions, nutritionally undemanding, and readily managed in industry. It is assumed, therefore, that they are also easy to maintain. However, a number of different maintenance methods commonly used result in poor survival levels and instability of properties.

Although the factors affecting survival are still not fully understood at the subcellular level, the poor performance of yeasts may be attributed in part to the large size of cells compared with bacteria and the absence of the kind of resistant spores produced by many of the higher fungi. In the light of present knowledge, therefore, high survival rates can best be achieved by careful attention to growth

MAINTENANCE OF MICROORGANISMS
ISBN 0 12 410350 2

conditions, choice of cryoprotectant, and the techniques used for preservation and revival.

Percentage survival of the total population levels following subculturing, drying or freeze-drying are generally low although 'cultures' may appear viable (Kirsop, 1974, 1978). By contrast, survival following storage in liquid nitrogen is high, often reaching levels of 100% (Hubalek and Kochova-Kratochvilova, 1978). There appears to be no relationship between survival and taxonomic position and the factors determining survival are specific for each strain. This means that a maintenance method that is satisfactory for one strain may be unsuitable for others, and generalizations on the effects of maintenance method on survival should be viewed with caution. If only a few yeasts are to be maintained, it may be possible to establish the optimum conditions for each strain; if larger numbers are involved, it may only be practicable to select the best method for most of the strains.

If strain stability is of paramount importance the choice of maintenance method becomes critical. Any method that enables cell division to occur should be rejected. It has been shown (Kirsop, 1974) that following freeze-drying yeasts in general remain un-changed morphologically, physiologically and with regard to industrial properties, although other authors have shown that genetic change can occur (Souzu, 1973). Again, some laboratories have found that yeasts dried on silica gel may show substantial changes (Bassel et al., 1977; Kirsop, 1978), whereas others report satisfactory results (Woods, 1976). It has been reported by Bassel et al. (1977) that genetically marked strains retain their characteristics after drying on filter paper and there is growing evidence that many strains stored in liquid nitrogen remain stable (Wellman and Stewart, 1973; Hubalek and Kochova-Kratochvilova, 1978).

Whichever method is selected the most suitable growth conditions for the cells must be established. It has been shown (Kirsop, 1978) that both the age of the culture and the oxygen availability during growth may substantially affect survival levels. Thus post-logarithmic cells almost always survive better than younger cells and better results with freeze-drying are obtained with oxygen limited growth using static cultures. By contrast, survival following storage in liquid nitrogen is generally better if cells are grown aerobically using shaken cultures.

It is clear that optimum survival depends upon many factors and that selection of the method to be used will rest not only upon the specific requirements of the strain to be maintained but also upon

Table I: Comparison of preservation methods

Method	Survival level (% population)	Shelf life (years)**	Stability	Convenience (rated 1, poor, to 5, good)	Economy
Subculturing:					
in broth	—	0.5	1	5	5
on agar slants	<10	1	1	5	5
on agar slants with oil overlay*	—	2	—	4	5
in water*	—	several	—	5	5
Drying:					
on filter paper* (genetic strains only)	—	3–6	5	4	5
on silica gel	—	1–5	4	4	5
Freeze-drying:	0·01–30	5–30	4	2	3
Freezing in liquid nitrogen:					
in ampoules	20–100	>5	4	3	4/5
in straws	20–100	>5	4	3	4/5

* Methods not used by NCYC.
** Known shelf life. These figures may be shown to be much higher as further information becomes available.

the availability of equipment, the experience of staff, and the purpose for which maintenance is required.

II. MAINTENANCE METHODS AVAILABLE

Ten methods suitable for the maintenance of yeast cultures are described. Most have been used by the National Collection of Yeast Cultures (NCYC) and information relating to survival level, shelf life, and the suitability of the method for different yeasts is known. Several other methods (marked with an asterisk) have not been used by the NCYC but are included as they may be more suitable in certain circumstances; with these methods references are given where possible, but in some cases the method has not been published. The methods are: subculturing in broth, on agar, on agar with oil (*), for strains with special requirements, in water (*); drying on silica gel and on filter paper (*); freeze-drying; and freezing in liquid nitrogen in ampoules or in straws.

Table I gives general information on the methods described and is intended only as a guide. As has been indicated, resistance to preservation is very strain specific and percentage survival and shelf-life figures are approximations. Convenience and cost will depend on facilities already available in laboratories.

III. SUBCULTURING

Subculturing has been used successfully for many years and the method continues to be useful, particularly in the short term. It is simple to do, quick to carry out, and relatively inexpensive. However, it is now recognized that substantial variation may take place in strains maintained in this way over a long period. It has been found, for example, that 10% of strains showed change with regard to flocculation behaviour following maintenance by subculturing over a period of ten years; other morphological and physiological properties were found to show variation to a greater or lesser extent (Kirsop, 1974).

If strain stability is of major significance, therefore, subculturing should be minimized.

A. *Subculturing in Broth*

1. Preparation of media
10 ml amounts of YM broth (Difco Laboratories, 0711–01) are

dispensed into screw-cap McCartney bottles. The medium is sterilized by autoclaving at 121°C for 15 min. Duplicate bottles are prepared for each culture, one labelled A and the other B (see section III.A.5).

2. Inoculation

A loopful from the B bottle of the old stock culture is transferred aseptically to each of the new bottles, the new B bottle being inoculated first.

3. Incubation

Inoculated bottles are incubated, without shaking, at 25°C for 72 h.

4. Viability estimation

Each bottle is examined visually for growth at 72 h and macroscopic characteristics (film, ring, colour and nature of deposit) are recorded. If little growth has occurred, cultures are reincubated for a further period and examined daily. It may be necessary to provide added aeration by shaking the cultures (for *Cryptococcus*, *Rhodotorula*, *Sporobolomyces* and other aerobic genera).

5. Storage

The duplicate cultures are stored at 4°C. During this time the A culture is used for all operations. The B culture is retained for the sole purpose of preparing the new stock cultures and should only be opened once.

6. Notes

Subculturing large numbers of strains is a tedious occupation and care should be taken to relieve monotony and minimize mistakes. It is sensible, for example, to place trays of old stock cultures and uninoculated bottles on opposite sides of the operator, and to transfer old and new bottles for each strain from the trays to a different, central position before starting to subculture.

7. Shelf life

The majority of yeast species, other than those with unusual cultural requirements, will remain viable under these conditions for 6 months and in all probability for a much longer period. In general, fermentative species survive better than non-fermentative species and some of the latter may need subculturing more frequently (perhaps at 2-monthly intervals).

Species of the genera *Brettanomyces* and *Dekkera* produce large amounts of organic acids that may reduce shelf life; these strains may need more frequent subculturing or may be maintained on a medium containing $CaCO_3$ to neutralize the acids.

It has been found that some yeast strains are sensitive to storage at 4°C. These strains must be stored at a higher temperature and, since growth will continue slowly, be subcultured more frequently.

8. Yeasts that can be maintained by subculturing in broth

The NCYC includes yeasts of most genera and all have been maintained by this method for periods of up to 60 years. The reservations made above in section III.A.7 apply, and it should be remembered that the method may result in considerable strain drift. Since the reasons for loss in viability are not clearly understood, strains should be monitored in the early periods of storage in order to determine which are more sensitive.

Genera most often found to contain strains requiring special treatment are listed in Table II, but it should be remembered that such strains may occur in other genera.

Table II: Genera containing strains requiring special treatment

Genus	Treatment
Brettanomyces Dekkera	More frequent subculturing; $CaCO_3$ in medium.
Bullera Sporobolomyces	More frequent subculturing; longer growth period before storage; aerobic growth.
Kloeckera Hanseniaspora	More frequent subculturing; added vitamins may be required.
Cryptococcus Lipomyces Rhodotorula	More frequent subculturing; aerobic growth.
Schizosaccharomyces	More frequent subculturing.

B. Subculturing on Agar Slants

1. Preparation of media

10 ml amounts of YM agar (Difco Laboratories, 0711–02) are dispensed into screw-cap McCartney bottles, autoclaved at 121°C for 15 min, and allowed to set at an inclined angle to form slants. All other information given in section III.A is applicable to this method.

2. Notes

Ascosporogenous strains may sporulate on agar slant cultures when stored for prolonged periods and this may lead to strain instability. If preservation of strain characteristics is a priority, this method should not be used. If no alternative method can be used, the yeast for subculturing should be taken from the bottom of the slant since ascospores are generally formed at the top.

3. Shelf life

Many yeasts will survive for longer periods on agar slants than in broth, particularly the non-fermentative genera (see section III.A.7).

4. Yeasts that can be maintained by subculturing on agar slants

Although this method has not been used by the NCYC for many years, all yeasts in the collection were maintained on agar slants in the past and records show a high recovery level. Non-fermentative strains often survive better on agar slants than in broth and, since many are anascosporogenous, the method may be particularly suitable for them.

C. Subculturing on Agar Slants with Oil Overlay

This method has not been used by the NCYC but has been recommended by a number of authors (see Onions, 1971) and has the advantage of extending the shelf life of agar slant cultures.

1. Preparation of medium

As for section III.B.1.

2. Preparation of oil

B.P. medicinal oil (BDH Chemicals Ltd) is dispensed into screw-cap bottles and sterilized by autoclaving at 121°C for 15 min. Some laboratories recommend a longer sterilization time, followed by drying in an oven for several hours. Rubber washers should be removed from bottles before use.

3. Inoculation

Using a wire loop, slants are inoculated from an actively growing culture.

4. Incubation

The culture is allowed to grow at 25°C for 72 h or until good growth is obtained.

5. Oil overlay

The sterile oil is transferred aseptically to the incubated slant culture so that the oil level is 1 cm above the top of the agar slant.

6. Storage
As for section III.A.5.

7. Notes
Care must be taken during subculturing not to let the inoculation loop splutter on flaming. Pathogens should not be maintained by this method.

8. Shelf life
The shelf life of yeasts maintained by method section III.B. may be substantially extended by covering slant cultures with a layer of sterile mineral oil and is generally in the order of 2 to 3 years.

9. Yeasts that can be maintained by subculturing on agar slants with oil overlay
The NCYC has no direct experience of this method, but it has been used by a fairly large number of laboratories over the years. There is little documentation regarding the yeast species that have been maintained successfully.

D. *Subculturing Strains with Special Properties*

Strains in which specific properties must be retained, and for which freeze-drying or freezing seem inappropriate, may be subcultured in medium that will select for the required characteristic. Thus strains resistant to specific inhibitors or strains tolerant of high sugar levels may be maintained in media containing appropriate levels of inhibitor or sugar. The number of selective media of this kind is unlimited, and shelf life and suitability of the method for different yeasts will be strain specific and must be predetermined. Methods described in sections III.A and III.B can usually be adapted to meet requirements of this kind.

E. *Subculturing in Water*

This method has not been used by the NCYC, but has been described by Odds (1976) and used successfully by him for a number of years.

1. Method
Growth from a late-logarithmic slant culture is suspended in sterile distilled water by agitation with a wire loop. The suspended culture is transferred aseptically to a sterile container so that 90% of the volume is filled with suspension. The culture is stored at room temperature.

2. Yeasts that can be maintained in water

Candida species and a few strains from *Saccharomyces*, *Cryptococcus*, *Trigonopsis*, *Rhodotorula* and *Schizosaccharomyces* have been preserved for more than 4 years by Odds (1976).

IV. DRYING

A. Paper Replica Method

This method has been developed by Bassel *et al.* (1977) and contributed by C. R. Contopoulou of the Yeast Genetics Stock Center, University of California, Berkeley, USA.

1. Preparation of paper replicas

Whatman No. 4. filter paper (Scientific Supplies Co. Ltd) is cut into small sections (approximately 1 cm^2) several of which (4–5) squares are placed on a piece of aluminium foil (7 cm × 6 cm); the foil is folded once and the packet autoclaved at 121°C for 15 min. One packet is used for each strain to be stored.

2. Inoculum

The cells to be stored are grown as a heavy patch on a plate containing a medium on which optimum growth can be attained (usually yeast extract–peptone–dextrose agar, YEPD). The plate is incubated for several days at an appropriate temperature (usually 30°C).

3. Suspending medium

Evaporated milk (any proprietary brand from food shop) is used as the suspending medium. Approximately 0.2 ml drops are transferred aseptically into sterile petri dishes (3–4 per plate).

4. Inoculation of paper replicas

As many cells as can be retained on the blunt end of a sterile toothpick are transferred from the inoculum plate to a drop of sterile milk and mixed thoroughly. Using toothpicks or sterile forceps, the sterile filter paper sections are immersed in the cell suspension and returned to the folded aluminium foil, leaving the remaining three edges unsealed to facilitate drying. Heavy suspensions of cells improve the chances of survival over longer periods of storage.

5. Drying

The packets are placed in a desiccator and allowed to dry for 2–3 weeks at 4°C. To prevent filter papers falling out from the packets and possibly causing contamination, several packets are held together securely with a paper clip.

6. Storage
Once dry, the packets are removed from the desiccator and the three unsealed edges are folded. The packets are then ready for storage in a dry container kept at 4°C.

7. Revival
Small pieces of the paper replicas are removed aseptically from the foil packets and are streaked across a plate containing the appropriate medium, leaving the filter paper in one corner of the plate. The plate is incubated at 30°C (23°C for temperature sensitive strains) for several days until good growth is obtained. If possible, individual clones are picked off to test for markers.

8. Shelf life
Cultures are revived and replaced at intervals of 2–3 years.

9. Yeasts that may be maintained by this method
The method has been used successfully for haploid and diploid strains of *Saccharomyces cerevisiae* with a wide variety of genetic markers. About 99% of several hundred strains have survived storage by this method for periods of 3–6 years.

Other species of *Saccharomyces* and other genera may not survive. Poor results have been obtained with strains of *Yarrowia lipolytica*.

B. Silica Gel Method

This is an adaptation of the method described in Ch. 11, section III.A. The method has been used successfully by C. F. Roberts of the Department of Genetics at Leicester University.

1. Preparation of silica gel
Purified silica gel (BDH Chemicals Ltd), mesh 6–22, is poured into McCartney bottles to a depth of about 1 cm. The gels are sterilized in an oven at 180°C for 90 min. The sterile gels are stored in a warm, dry atmosphere.

2. Suspending medium
A 5% skimmed milk solution (Oxoid Ltd) is prepared in distilled water. The solution is distributed in approximately 10 ml amounts in McCartney bottles, sterilized by autoclaving at 116°C for 10 min, and stored at 4°C.

3. Inoculum
The yeast culture is grown on a YM agar (Difco Laboratories,

0711–02) slant culture for 72 h at 25°C.

4. Inoculation of gels

Gels and milk solution are placed in a refrigerator for 24 h before use to become cold. The cold gels are transferred to an ice tray for inoculation. Using a Pasteur pipette yeast cells are washed off the agar slant culture with the cold milk and a few drops of the yeast suspension are transferred to each gel. The inoculated gels are shaken to disperse the cells and are returned to the ice tray where they remain for a further 30 min.

5. Drying

The gels are kept at room temperature for about 2 weeks to dry, care being taken to screw the caps tightly.

6. Storage

After 2 weeks, when the gel crystals separate readily, the gel bottles are transferred to an air-tight plastic container in the bottom of which is a layer of indicator silica gel (BDH Chemicals Ltd). The lids are put in place and the containers stored at 4°C. The indicator gel is checked from time to time and replenished or redried by heating in an oven at 180°C for 2 h, as required.

7. Revival

A few crystals are either shaken onto YM agar plates (Difco Laboratories, 0711–02) or into YM broth (Difco Laboratories, 0711–01) and incubated at 25°C for about 3 days, depending on the growth rate of the strain.

8. Shelf life

Viability using this method is strain specific and the NCYC has found some strains to be dead after 3 months and others still to be alive after 2 years' storage. Woods (1976) has recovered strains after 5 years' storage.

9. Notes

 (i) It is important to keep all apparatus very cold during inoculation to minimize the effects of the heat generated when the gels are hydrated.
 (ii) The gels should not be saturated with yeast suspension; 2 or 3 drops to each gel are sufficient.

10. Yeasts that may be maintained by storage on silica gel

This method has been used successfully by some workers and unsuccessfully by others. A summary of these findings is given below.

Of 25 species (representing 17 genera) preserved by the NCYC 50% could be recovered after storage for 2 years; others were no longer viable after 1 week. Survival was strain specific and could not be related to taxonomic position. Some strains of *Saccharomyces cerevisiae* that survived were no longer typical with regard to fermentation characteristics.

Woods (1976) found that purine-requiring mutants and polyene-resistant mutants of *Saccharomyces cerevisiae* survived unchanged for 5 years; this laboratory was less successful with strains of *Candida albicans* and *Candida tropicalis*, which confirms the experience of the NCYC.

Bassel *et al.* (1977) report inconsistent results after storage of *Saccharomyces cerevisiae* strains for a few years. Particularly sensitive were temperature-sensitive lethals, respiratory-deficient mutants, fatty acid-requiring strains and certain other auxotrophic mutants. They now use the method in section IV.A. in preference.

The NCYC has found that some strains will survive on one occasion but not on another, and it is felt that survival could be improved with further research into the effects of different parameters. The method does not offer any advantage to the NCYC, which is required to maintain a very wide spread of microorganisms, but may well be useful for more specific purposes.

V. FREEZE-DRYING

This is a two-stage method using a centrifugal freeze-dryer, Edwards Model 2A/110 or EF03 (Edwards High Vacuum). It can be used equally well with other centrifugal machines.

A. Centrifugal Freeze-drying

1. Preparation of ampoules
Glass ampoules (FBG-Trident Ltd) are washed in detergent and rinsed in demineralized water before use. Labels for the ampoules are prepared from strips of Whatman's No. 1. filter paper (Scientific Supplies Co. Ltd). The number of the yeast and the date are printed on the label either in pencil or using a stamp with ENM quick-drying, non-toxic ink (Rexel, obtainable from stationers). The numbers are printed as near to one end of the label as possible.

The label is folded in half lengthwise (see Fig. 1) and inserted in the tube so that the writing faces outwards. The tubes are loosely plugged with non-absorbent cotton wool and those for each yeast are

Fig. 1: Printed label for insertion in ampoule.

placed in a separate tin. The tins are either sterilized in an oven or autoclaved at 121°C for 15 min. After autoclaving they are dried in an oven.

2. Inoculum
The culture to be freeze-dried is grown without aeration in YM broth (Difco Laboratories, 0711–01) at 25°C for 72 h, the amount needed depending upon the number of ampoules to be prepared. Each ampoule requires 0.1 ml of a suspension containing at least 10^6 cells ml^{-1}; in general, increasing the number of cells in the inoculum increases proportionately the number of surviving cells.

3. Suspending medium
Glucose, to a final concentration of 7.5%, is dissolved in inactivated horse serum No. 5 (Wellcome Reagents Ltd) and the mixture is sterilized by filtration and stored in McCartney bottles in a refrigerator at 4°C; sucrose or inositol may be used in place of glucose; inositol is known to provide a longer shelf life for bacteria. Other possible protective substances include skimmed milk (20%) and sodium glutamate (10%).

4. Inoculation of ampoules
Equal amounts of inoculum and suspending medium are mixed aseptically in a sterile bottle, the amount prepared depending on the number of ampoules to be freeze-dried. The inoculum is not washed before use. A sterile 30-dropper Pasteur pipette is used to transfer 6 drops (0.2 ml) of the mixture to each ampoule, after which the cotton-wool plug is replaced. Care is taken not to touch the sides of the ampoule with the pipette, as this will distort survival levels. The original culture is retained for a viability estimation. If this is not to be done immediately, the culture is put into the refrigerator to prevent further cell division.

5. Primary freeze-drying
The inoculated ampoules are placed immediately in the centrifuge head of the freeze-dryer with the writing on the labels facing towards the centre of the centrifuge head, so as not to be obscured by the dried yeast after centrifugation. The centrifuge head is placed on the

spindle in the drying chamber. If the freeze-dryer uses phosphorus pentoxide as a desiccant the four trays are filled in a fume cupboard and placed in the drying chamber. *It is important that protective clothing is worn when handling phosphorus pentoxide.* The centrifuge motor and the vacuum are switched on and primary drying is continued for 3 h. At the end of this time, the centrifuge head is removed from the chamber. The yeast in the ampoules should appear completely dry.

6. Ampoule constriction

The projecting ends of the cotton-wool plugs are trimmed and the remainder of the plugs pushed halfway down the ampoules with a glass rod. The ampoules are then constricted above the level of the plug either by hand or using an ampoule constricter (Edwards High Vacuum). The phosphorus pentoxide in the freeze-dryer is re-plenished with fresh desiccant and returned to the chamber ready for secondary drying.

7. Secondary drying

The chamber is closed and the manifold is put in place. The constricted ampoules are placed on the manifold. Spare ampoules are used to fill any spaces. Secondary drying is generally continued overnight. Alternatively, the secondary drying may be terminated at 2 h, allowing the whole operation to be completed in one working day. After this time the ampoules are sealed under vacuum using a Flaminaire blow torch (Longs Ltd). Details of the run and the number of satisfactory ampoules obtained for each yeast are recorded. One ampoule of each culture is retained for a viability estimation. The spent P_2O_5 is covered with water *in a fume cupboard* and discarded. If no fume cupboard is available the trays can be left in a safe place for the desiccant to deliquesce. The trays are washed thoroughly and left in a warm place to dry.

8. Revival

Ampoules are opened as in 9(b) below using YM broth (Difco Laboratories Ltd) to resuspend the cells. Attempts to improve survival levels by using different suspending media, times and temperatures have not been very effective although the use of half-strength YM broth generally gave higher viable counts.

9. Viability counts

(a) On cultures before freeze-drying. 1 ml of the original suspension is added to 9 ml sterile glass-distilled water. Further dilutions to 10^{-7} are made. 3 drops from a 30-dropper pipette (0.1 ml) of

dilutions 10^{-2} to 10^{-7} are placed onto YM agar (Difco Laboratories Ltd, 0711–02), according to the Miles and Misra (1958) plate count method. Plates are incubated at 25°C for 72 h, or longer if necessary, care being taken to keep the plates horizontal so that the drops remain discreet. Drops containing 10 to 30 colonies are used for estimating viability. The number of cells ml^{-1} inoculated into the ampoule = no. of colonies in 3 drops $\times$ 10 $\times$ dilution factor.

(b) On freeze-dried cultures. This count is carried out as soon as possible after freeze-drying. The ampoule is marked with a file at or just above the level of the cotton-wool plug. A molten glass rod is applied to crack the glass. The tip of the ampoule is removed and placed in a container for later sterilization.

1.0 ml of YM broth is put into a sterile bijoux bottle. Using a Pasteur pipette some of the YM broth from the bijoux is added to the ampoule. The yeast is thoroughly resuspended and the whole of the suspension returned to the remaining YM broth. Using sterile forceps the label is removed from the ampoule and placed in the bijoux. A further few drops of YM broth from the bijoux may be used to wash out the ampoule again, the drops being returned to the bijoux. The suspension is a 10^{-1} dilution. Of this, 0.5 ml are transferred to 4.5 ml sterile water (10^{-2}) and further dilutions to 10^{-6} are made, plated, incubated and counted as in (i). The percentage viability of the freeze-dried culture is calculated and recorded.

10. Storage
Ampoules are stored at 4°C in the dark. Each ampoule is tested for good vacuum before storage using a high frequency spark tester (Edwards High Vacuum).

11. Notes
The NCYC has found that the degree of oxygenation provided during growth of the inoculum may affect survival. Although response to oxygen is strain specific, in general, higher survival levels are obtained following growth with restricted access to oxygen. It was also found that growth on nutritionally poor medium before freeze-drying gave higher survival levels with sensitive strains.

12. Shelf-life
Survival levels of yeasts are generally low (see section V.A.13), but losses during storage are minimal. Although cultures have been preserved by the NCYC for over 30 years, and for longer periods by other laboratories, occasional strains suddenly lose viability. It is

important, therefore, to monitor viability on a regular basis.

13. Yeasts that can be maintained by freeze-drying

Survival of yeasts following freeze-drying is remarkably strain specific and generalizations regarding survival levels should be viewed with caution. Nevertheless, all cultures maintained at the NCYC, and covering nearly all yeast genera, have been recovered successfully, although the percentage survival of the population is low. Thus, the average viability figures obtained at NCYC for the genus *Saccharomyces* is 5%, for *Candida* 13%, and for *Brettanomyces* 2%. The genera in which survival levels have been particularly low are: *Brettanomyces, Dekkera, Bullera, Sporobolomyces, Leucosporidium, Rhodosporidium* and occasional strains of almost all other genera.

Apart from increased levels of respiratory deficient mutants, the NCYC has detected little change in the characteristics of freeze-dried cultures and the method has advantages with regard to strain stability.

VI. FREEZING

A. *Freezing in Liquid Nitrogen (Ampoules)*

This is a two-stage method in which cells are cooled at an uncontrolled rate to $-30°C$, allowed to dehydrate at this temperature and cooled, again at an uncontrolled rate, to $-196°C$. It has the advantage that controlled freezing rate equipment is not required.

1. Preparation of ampoules

Plastic screw-cap 2 ml ampoules are supplied sterilized from the manufacturer (Nunc, Gibco-Europe Ltd). They are labelled with the yeast number and date using a black Pentel pen (Scientific Supplies Co. Ltd), other colours having been found less satisfactory.

2. Inoculum

The culture to be frozen is grown in YM broth (Difco Laboratories Ltd, 0711–01) at 25°C for 72 h on a reciprocal shaker. Each ampoule requires 0.5 ml of a suspension containing between 10^6 and 10^7 cells ml^{-1}. Cell concentration has been found to have little effect on the percentage of cells surviving.

3. Preparation of cryoprotectant

A 10% glycerol solution is prepared, filter-sterilized, and stored in sterile screw-cap bottles (see note 9(a) for other suitable cryoprotectants).

4. Inoculation of ampoules

Equal amounts of inoculum and cryoprotectant are mixed asepti-cally in a sterile bottle. The final concentration of glycerol is thus 5%. One ml of the mixture is transferred to each of the sterile ampoules. Care is taken that the washer is not displaced when the cap is screwed tight as this may lead to leakage of liquid nitrogen into the ampoule; it should only be tightened until initial resistance is felt.

5. Freezing

(a) Primary freezing. It is very important to wear protective clothing when using liquid nitrogen refrigerators or handling frozen specimens. Ampoules are frozen to −30°C by placing in a refrigerated room. If a room at −30°C is not available a refrigerated cabinet or cooling bath (Camlab Ltd) may be used. If aluminium canes (Union Carbide) are to be used for the secondary freezing, ampoules may be placed on the canes at this stage. If secondary freezing is to take place in storage drawers (Union Carbide), however, the ampoules may be well-spaced in wire racks for the primary freezing. The cooling rate is not critical in this method, but is probably in the region of 5°C min^{-1}, depending upon such factors as the size of samples and containers. Cells are allowed to dehydrate at −30°C for 2 h.

(b) Secondary freezing. The ampoules are transferred to the canisters or storage drawers of the refrigerator (Union Carbide) and immersed in the liquid nitrogen, care being taken to prevent the samples from thawing. If the distance between primary and secondary freezing containers is great, samples should be transported in a chilled Dewar.

6. Revival

Cultures are thawed rapidly by transferring ampoules to a water bath at 35°C and agitating them until completely thawed.

7. Viability counts

Viability counts are carried out, with appropriate adjustments, as described in section V.A.9.

8. Storage

Ampoules are stored in the liquid nitrogen refrigerator (Union Carbide), care being taken to maintain the liquid nitrogen at such a level that ampoules are completely submerged (see note 9(c) below).

9. Notes

(a) A number of other cryoprotectants have been used successfully

both by the NCYC and other workers. Some have been used for a fairly wide range of yeasts, others with a few strains only. Substances used successfully include glycerol (20%, 10%, 5%), dimethyl sulphoxide (DMSO, 10%), glycerol plus DMSO, ethanol (10%), methanol (10%), YM broth and hydroxyethyl starch (10%, 5%).

(b) The NCYC has found that primary freezing and dehydration at −20°C, −30°C or −40°C for 1, 2 or 3 h is equally successful for the two test strains of *Saccharomyces cerevisiae* used to develop the method. The intermediate protocol (−30°C for 2 h) has been adopted and proved successful, so far for all species.

(c) Cultures have been stored successfully in the vapour phase of liquid nitrogen by other workers, but as it is generally accepted that biochemical and biophysical processes may still take place, albeit slowly, at temperatures above −139°C, the NCYC prefers to store cultures at −196°C. For shorter storage periods the higher temperatures may be considered adequate.

(d) The NCYC has found that, in general, higher levels of survival are obtained from aerobically grown cultures than from those grown with limited access to oxygen. This should be compared with growth conditions found to give optimum survival in freeze-drying (see sections V.A.2 and V.A.11).

(e) Although the NCYC has experienced no leakage of liquid nitrogen into filled ampoules, other workers have evidence that this may occur. It can be overcome by (1) storing in the vapour phase (but see VI.A.9.c. above) (2) using the straw method (see section VI.B. below) or (3) placing the ampoules in polypropylene sleeves (Union Carbide).

10. Shelf life
The long-term survival of yeasts in liquid nitrogen is not well documented, but all evidence suggests that losses during storage are very slight. The NCYC has detected no drop in viability in test strains over a period of 26 months and others record good survival for periods of up to 4 years. In view of the high initial survival rates, shelf life can be expected to be good.

11. Yeasts that can be maintained by freezing in liquid nitrogen
All yeast strains that have been preserved by the NCYC using this method show high viable counts. Although survival is strain specific, the average survival level for all genera immediately after freezing is greater than 60%. Average survival levels are for the genus *Saccharomyces* 65%, for *Candida* 73% and for *Brettanomyces* 64%.

Strains from the following genera have been successfully preserved by the NCYC:

Brettanomyces	*Metschnikowia*	*Saccharomycodes*
Bullera	*Nematospora*	*Schizosaccharomyces*
Candida	*Pachysolen*	*Sporobolomyces*
Citeromyces	*Phaffia*	*Torulaspora*
Cryptococcus	*Pichia*	*Trigonopsis*
Hansenula	*Rhodosporidium*	*Wingea*
Kloeckera	*Rhodotorula*	*Yarrowia*
Kluyveromyces	*Saccharomyces*	*Zygosaccharomyces*
Lipomyces		

B. *Freezing in Liquid Nitrogen (Straws)*

This is a miniaturized adaptation of the method described in section

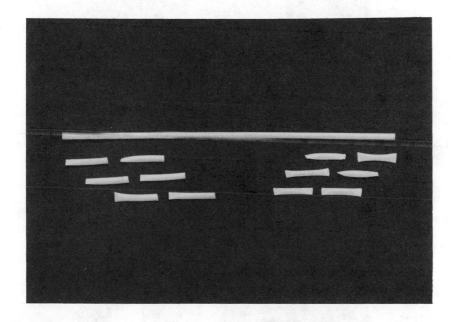

Fig. 2: Polypropylene straws—uncut (top); sealed at one end (left); sealed at both ends (right).

VI.A above. It has two advantages. First, it provides a considerable saving in storage space and, secondly, it provides additional security against possible contamination through leakage of liquid nitrogen into ampoules that are stored in the liquid phase. Storage in straws was first described by Gilmour *et al.* (1978) using artificial insemination straws; adaptations of the method using different kinds of straws are in use in a number of laboratories.

1. Preparation of straws

Coloured polypropylene straws (Sweetheart International Ltd) are cut into 2.5 cm lengths. One end of a straw is sealed by holding firmly in a pair of unridged forceps 1 mm inwards so that the projecting end is 1 cm from the flame of a fish-tail bunsen burner. The polypropylene melts almost immediately and forms a strong seal that sets firm within a second or two.

Fig. 3: Sealing polypropylene straws.

The straws are placed in a glass petri dish and autoclaved at 121°C for 15 min. The different coloured straws may be used for colour-coding yeast strains.

2. Inoculum
The inoculum is prepared as in section VI.A.2.

3. Preparation of cryoprotectant
The cryoprotectant is prepared as in section VI.A.3.

4. Inoculation of straws
Equal quantities of inoculum and cryoprotectant are mixed aseptically in a sterile bottle. A single straw is removed with forceps from the petri dish and filled with inoculum using a Pasteur pipette. When filling it is necessary to place the end of the pipette close to the sealed end of the straw and to fill to within 3 mm of the open end. Each straw will hold about 0.1 ml of inoculum. The filled straw is then sealed at the open end as described in section VI.B.1 above.

Six straws are placed in each plastic, screw-capped 2 ml ampoule (Nunc, Gibco-Europe Ltd) of the kind used in section VI.A.

5. Freezing
Primary and secondary freezing are carried out as in section VI.A.5 above. Straws are frozen in the ampoules; this is easier than freezing the straws separately and transferring them, frozen, to the ampoules. No difference in survival level has been detected between straws frozen individually or in ampoules.

6. Revival
Using sterile forceps, a straw is removed from the ampoule and placed immediately into a screw-cap bottle containing water at 35°C. The bottle is shaken to facilitate rapid thawing. Several straws may be thawed simultaneously in this way.

7. Viable counts
Before opening the straws, the cells are resuspended by squeezing the straws several times. The straws are then wiped with 95% alcohol and one end is cut off using sterile scissors. The contents are removed using a Pasteur pipette. It may be necessary to disperse cells further by repeated pipetting at this stage.

Two drops of the suspension (0.06 ml) are transferred to 0.54 ml sterile water to make a 10^{-1} dilution. Further dilutions, plating and counting are carried out as in section V.A.9.

8. Storage
Ampoules are stored as in section VI.A.8.

9. Notes
(a) Notes (a), (b), (c) and (d) in section VI.A.9 apply also to this method. It is important that the outside of the straws is kept dry during filling as wet straws do not seal well. When filling several straws with the same inoculum it is convenient to lay each straw as it is filled against a glass rod in a sterile petri dish until all straws are filled. The straws are then sealed and put into ampoules. This is more convenient than filling and sealing each straw separately.

If the outsides of the straws are dry, they do not adhere to each other when placed in ampoules. Removal of straws from ampoules is facilitated if straws vary slightly in length. The different coloured straws can be used to colour code yeast strains to aid retrieval from the refrigerators, and it is clearly sensible to store only one yeast strain in each ampoule.

(b) Other makes of straw have been used successfully in other laboratories. The straws used for the storage of semen at artificial insemination centres have been found particularly suitable (Instruments de Médecine Vétérinaire).

10. Shelf life
The NCYC has only been using this method since 1982. No loss of viability has been noticed in the test straws during this time and the high survival levels would suggest good shelf life.

11. Yeasts that can be maintained by storage in straws
Results obtained so far at the NCYC show that survival levels are equal to (and sometimes higher than) those obtained in method VI.A. These results suggest that species that can be frozen in ampoules can be frozen equally successfully in straws.

13

Maintenance of Algae and Protozoa

E. A. LEESON, J. P. CANN and G. J. MORRIS

Institute of Terrestrial Ecology
(Natural Environment Research Council)
Culture Centre of Algae and Protozoa
Cambridge, UK

MAINTENANCE OF MICROORGANISMS
ISBN 0 12 410350 2

I. SUBCULTURING ALGAE

A. *Introduction*

Subculturing is the transfer of viable material to fresh growth medium. Although a simple method, it has the disadvantage that strains are subjected to continual selection pressures whilst being maintained in an artificial environment. During long-term culture, this may result in changes in some strains. Subculturing (Belcher and Swale, 1982) may be carried out using plugged sterile pipettes or a microbiological loop, using aseptic techniques where necessary. Media should be decanted into clean vessels which are then plugged or capped and sterilized. When handling many strains of algae, practical considerations seldom permit each culture to be kept under individual optimum conditions. Individual strains may show differing requirements for light, temperature and pH. Types of nutrition found within this group range from obligate phototrophy in pigmented forms to heterotrophy in colourless forms. Nutritional requirements vary not only between groups but also from strain to strain, and extensive trials are required to find the best medium for individual algae. However, healthy growth, adequate for the continuation of an algal strain, can usually be achieved using one or more of the three basic types of algal culture media. These are: (a) biphasic culture (soil and water); (b) liquid media; (c) solidified agar plates or slopes.

A number of factors must be taken into account when choosing a suitable culture medium. For example, the status of the alga with regard to other microorganisms is important. When grown on mineral media algae may benefit from nutrients made available by other microorganisms in the culture. Enhancement of algal growth may be obtained by the addition of extra organic substrate to the media. Soil or plant material extracts are commonly used for this purpose as the organic content is beneficial to the alga but insufficiently high to encourage excessive bacterial growth. It can be generally assumed that an organic content of more than 0.1% is suitable only for strains which have been isolated into bacteria-free culture.

Many commonly cultured algal strains are not strictly photo-autotrophic and are able to use, or may show a specific requirement for, external organic carbon sources. Some euglenoid flagellates, several cryptophytes, and many volvocalean flagellates use acetate (Hutner and Provasoli, 1951), while others such as some blue green algae, diatoms and chlorococcales may use sugars. Simple organic acids and alcohols are also utilized by some species (Venkataraman, 1969).

Axenic cultures may benefit from the addition of vitamins which may be supplied in the form of yeast extract. In addition, media may be supplemented with nitrogen. Nitrate is the source most commonly used, although ammonia, urea, proteose peptones and casamino acids may also be suitable. Nitrogen may be omitted from medium used in the culture of nitrogen-fixing blue-green algae (Cyanobacteria). This encourages the formation of heterocysts which may be a useful diagnostic feature of some strains.

When culturing marine strains, media based on filtered natural seawater or artificial sea water may be used. These media may be enriched (Stein, 1973).

Light is required by photosynthetic algae. Fluorescent tubes and incandescent bulbs provide a suitable light source, as does daylight in many cases. Care must be taken to avoid excessive localized overheating, either from over proximity to the light source or by placing cultures in direct sunlight.

Specific temperature requirements are variable between strains. Most freshwater algae will grow well at about 20°C, although some cyanobacteria prefer a slightly higher temperature range of about 20–25°C (Stein, 1973). Following the establishment of newly inoculated cultures, management of a collection may be facilitated by storage of cultures at temperatures and light intensities lower than normal. This will allow the continuation of adequate culture growth for strain maintenance but at a slower growth rate, thus reducing the number of subculturing manipulations required. A room illuminated by a north-facing window at approximately 15°C is suitable.

B. Biphasic Media

Biphasic or soil + water cultures are made by the partial sterilization of soil at the base of a vessel of water. Approximately 1 cm of soil (preferably of the garden loam type) is placed at the base of a suitable vessel, such as a beaker or tube, which is then filled with water and capped or plugged. The medium is steamed for 1 h and then allowed to cool. Steaming is repeated on two successive days. Most fungi and algal species are killed by this treatment but bacterial spores remain viable. This type of medium has the advantage of providing a buffered system in which trace elements are available in chelated form, minerals are present and other nutrients are supplied by the bacterial flora present (Pringsheim, 1946). These media are useful for culturing nearly all algae, particularly filamentous forms. Variations of the basic media may be made by the addition of cereal grains to the soil for supporting colourless species or small amounts of compounds such as $CaCO_3$ or NH_4MgPO_4, catering for more specific needs of members of the Chlorophyceae or Euglenophyceae respectively. The shelf life of these cultures may often be as long as several years, although more frequent checking of cultures is advisable.

C. Liquid Media

Many recipes for defined and semi-defined media suitable for algal cultures are described in the literature (Stein, 1973; Venkataraman, 1969; Rechcigl Jr., 1978) ranging from mineral media such as Bold's Basal Medium (Nichols and Bold, 1965) to media supporting heterotrophic forms such as *Euglena gracilis* medium (Asher and Spalding, 1982). For convenience, concentrated solutions of components may be kept as stocks. Media may be dispensed into suitable clean vessels such as test tubes or conical flasks, which are then fitted with a cotton wool plug or cap and sterilized by autoclaving at 115°C for 20 min. Liquid media such as Bold's Basal Medium enriched with 5% soil extract and soil extract media with added mineral salts (Asher and Spalding, 1982) are very useful for establishing many algae in culture. The former gives particularly good results with flagellates such as the Volvocales and other Chlorophyceae (Desmids and Chlorococcales). The soil extract medium supports healthy growth of most algal types including filamentous and planktonic Cyanophyceae. The shelf life of liquid cultures varies with each strain and ranges from 2 weeks for some flagellates to approximately 3 months for slower growing forms, such as some filamentous blue-green algae.

D. Agar Slopes

Slopes are prepared by the solidification of liquid media with approximately 1% agar. Agar powder should be dissolved in media prior to dispensing into test tubes, or other suitable vessels, which are
then capped or plugged and sterilized by autoclaving at 115°C for 20 min. Tubes should be cooled and allowed to set at an angle of approximately 30° to ensure the formation of an adequate surface on which to grow the culture. This type of culture is very useful for keeping long-term stocks of strains. The shelf life of these cultures

again is strain and media-dependent, but most cultures can be stored for periods of 1–6 months in this form. Plugs or caps should be well fitting but not totally sealed to the vessel, as gaseous exchange with the external environment is generally required. Slopes and plates can prove very useful in monitoring unwanted microbial contamination in axenic cultures. When using enriched media for this purpose contaminating fungi and bacterial organisms can often be detected by eye.

II. STORAGE OF RESISTANT STAGES IN THE LIFE CYCLE

Many algae tolerate adverse conditions in their natural environment by the formation of resistant stages such as zygotes or cysts. If these forms can be induced in the laboratory, they may be stored in a dry state and regerminated when required. Their longevity is variable. Zygotes of *Hydrodictyon africanum* have been successfully stored for 6 years and cysts of *Haematococcus pluvialis* have remained viable after 27 years in air-dried soil.

III. SUBCULTURING PROTOZOA

A. *Introduction*

Virtually all protozoan cultures can be maintained by subculturing in media with either bacteria or some other food organism present. Sometimes, however, it is more convenient to have a bacteria-free (axenic) culture growing in a rich sterile medium. Initial experiments on a range of protozoa have indicated that preservation in liquid nitrogen is difficult, with very low survival rates. Therefore, at the present time, routine subculturing appears to be the most reliable method for maintaining protozoa.

   The object of maintaining a culture is always to have a sufficient quantity of viable cells for subculturing into fresh media. Although most strains are fairly easily maintained, some are susceptible to very slight changes in temperature and pH. Before inoculating, media should stand for at least a day to reach gaseous equilibrium with the surrounding atmosphere, but can be kept longer for storage purposes under refrigerated conditions. When subculturing, one of the oldest tubes or beakers from stock is selected and checked for the presence of viable cells. With larger ciliates a hand lens is usually suitable for this purpose, but for smaller organisms the tube should be examined under a stereoscopic microscope. Alternatively, a drop of the culture can be placed on a slide and observed using a standard bright-field microscope. When the presence of viable cells has been established, the necks of both the old and new tubes are flamed and the old culture is carefully poured into the fresh medium. Alternatively, sterile plugged Pasteur pipettes are used to transfer cells using sterile technique.

After subculturing, the cultures are usually kept at room temperature (20°C) for a week until logarithmic growth phase is reached. They are then transferred to a cold room (15°C) for longer storage. The frequency of subculturing depends on the organism and the type of medium used. Media for maintaining protozoa range from autoclaved river water to complex agars and references and details of media follow.

B. Biphasic (Soil + Water) Media

With this method (Page, 1981; Asher and Spalding, 1982), the most useful type of soil to use is a good calcareous garden loam, but a range of different types of soil can be tried. Before autoclaving, a few barley grains or some potato starch are added. Starch is placed under the soil, but barley grains are left lying on top; the tubes are plugged before autoclaving or steaming for 1 h on each of 2 successive days. Beakers containing biphasic medium are covered with the lid of a glass petri dish or greaseproof paper to prevent drying out and contamination.

The following organisms can be maintained on biphasic media; food organisms are usually present and are carried over when subculturing:

Anthophysa (colonial flagellate)	*Colpoda*	*Paramecium*
Astasia	*Distigma*	*Polytoma*
Bodo	*Gyropaigne*	*Polytomella*
Chilomonas	*Hyalophacus*	*Rhabdomonas*
Coleps	*Menoidium*	*Spirostomum*
Colpidium	*Monas*	*Stentor*
		Urocentrum

C. Axenic Culture in Liquid Media

These are generally more complex media, although sterile E + S liquid (soil extract + salts) is a useful, easily prepared medium (Page, 1981; Asher and Spalding, 1982). Sterile technique is obviously important when handling axenic cultures. Organisms maintained in sterile media are: *Acanthamoeba, Astasia, Chilomonas, Euglena, Polytoma, Polytomella, Peranema* and *Tetrahymena*.

D. In Liquid Media + Food Organisms

A range of different liquid media are available for protozoa (Page, 1976; Page, 1981; Jeon, 1973). Usually the liquid (river water, Prescott and James's (P & J), Cerophyl-Prescott (CP), E + S) is sterilized in convenient sized flasks before use. When maintaining cultures of larger amoebae, such as *Amoeba proteus*, it is not necessary to autoclave the medium (P & J) before use. Protozoa are maintained as described below.

1. Amoeba proteus
The culture is grown in 7.5–9 cm diameter glass containing P & J medium plus 2 or 3 rice grains. *Chilomonas* is added to new isolates as a food source and is carried over on subculturing into new dishes. The amoebae are subcultured by pipetting once a month from the oldest viable culture, covering the new dish with "Clingfilm" and a glass petri dish lid to deter mites.

2. Dileptus
The culture is grown in P & J liquid medium. It is fed twice a week with *Colpidium*.

3. Heliozoans (*Actinophrys, Actinosphaerium, Raphidiophrys*) and *Discophrya*
These organisms are grown in smaller dishes containing autoclaved river water to which a chelating agent has been added (*Actinosphaerium* medium). They should be subcultured once a week by pipette and fed twice a week on *Colpidium*. About 3 ml of *Colpidium* are added to each dish with a 2-day interval between feeds. Many cultures will have accompanying bacteria as food source and boiled wheat or barley grains are generally added as a source of starch. Grass seed (untreated) is often a useful addition. Some organisms (e.g. *Nassula*) feed on filamentous blue-green algae such as *Phormidium* which should be sonicated before being used as a food source. The following protozoa are maintained in liquid culture plus food organisms:

> *Arcella* (no grains needed) CP liquid
> *Heteromita* CP liquid
> *Nucleosphaerium* (Filose amoeba) P & J + sonicated *Phormidium*
> *Opisthonecta* E + S + boiled wheat grain
> *Urocentrum* E + S + boiled wheat grain
> *Euplotes* E + S + rice grains (marine strains in sterile seawater)

E. In Liquid Overlay on Non-nutrient Agar

CP liquid and P & J liquid are suitable media for maintaining *Trinema* and *Euglypha* respectively in an overlay on non-nutrient agar in 4.5 cm diameter glass covered dishes. The protozoa are observed under a stereoscopic microscope before subculturing and transferred by pipette into new dishes every two weeks. (See Asher and Spalding, 1982).

F. On Agar Plates and Slopes

(See Page, 1976: Asher and Spalding, 1982.) Smaller amoebae are maintained on agar plates or, if cyst-forming, on agar slopes and stored at 8°C in universal glass containers. Cyst-forming amoebae should be excysted regularly to produce trophic stages, which in turn produce younger cysts for longer term storage. Amoebae such as *Acanthamoeba, Paratetramitus*, and *Vahlkampfia* need to be excysted only once a year. The amoeboid flagellate *Naegleria* should be excysted every 6 months. If no bacteria are present in the culture, plates of non-nutrient agar are streaked with the bacterium *E. coli* to provide a food source for the amoebae after excystment. Encysting amoebae are transferred by cutting a block of agar from a stock slope using a flamed surgical blade. The block is carefully inverted and placed culture side down over the streak of bacteria. A few drops of sterile amoeba saline are added to promote excystment, and the plates are left for one week. Amoebae are observed by inverting the petri dish under a ×5 objective and marking their positions with a felt-tipped pen. New agar slopes are streaked with *E. coli* and blocks cut from the plate using a flamed surgical blade are transferred to the new stock slopes. Amoebae which are not cyst-forming, such as *Saccamoeba limax*, can be maintained on plates at room temperature. Blocks are cut from an old stock plate in the way described above and new plates grown for stock.

If bacteria are present in the culture, an agar medium such as Cerophyll-Prescott agar can be used. The method of subculturing is as above, but no bacteria need be added to the new plates and slopes.

It should be borne in mind that some genera of common, free-living amoebae, such as *Acanthamoeba* and *Naegleria*, have been shown to be pathogenic to man. Possible routes of infection, such as the nose, eyes or mouth, should therefore be protected. This may require the use of a biological containment cabinet. If possible, the production of aerosols should be avoided when subculturing and

Table I: Published methods for the lyophilization of algae.

Species	Suspending Medium	Recovery (%) after 24 h	Reference
Bracteacoccus cinnabarinus	Horse serum	0.25	Daily and McGuire (1954)
Chlamydomonas pseudococcum	Horse serum	0.013	Daily and McGuire (1954)
Chlorella pyrenoidosa	Skim milk (10%) + monosodium glutamate (1%)	5	Tsura (1973)
Chlorella pyrenoidosa	Skim milk (10%)	5.3×10^{-3}	Holm-Hansen (1967)
Chlorella sp.	Skim milk (10%)	1.3	Holm-Hansen (1967)
Chlorophyta	Skim milk (10%)	39/106 strains viable; recovery from 10^{-1} to 10^{-6}%	McGrath et al. (1978)
Chlorophyta	Horse serum	15/22 strains viable	Daily and McGuire (1954)
Cyanophyta (7 strains)	Horse serum	All strains viable	Daily and McGuire (1954)
Cyanophyta (13 strains)	Lamb serum	All strains viable	Corbett and Parker (1976)
Nitzschia closterium	Skim milk (10%) + monosodium glutamate (1%)	0.5	Tsura (1973)
Phaeodactylum tricornutum	Skim milk (10%) + monosodium glutamate (1%)	1.5	Tsura (1973)
Scenedesmus obliqus	Horse serum	0.025	Daily and McGuire (1954)
Stichococcus bacillaris	Skim milk (10%)	3×10^{-1}	Holm-Hansen (1967)

used glassware should be soaked in strong disinfectant before being autoclaved.

IV. FREEZE-DRYING

Recovery of several species of algae has been reported after 24 h in the freeze-dried state (Table I). However, reported survival levels were extremely low, usually below 1%; this low level of cellular viability decreased further with prolonged storage (Holm-Hansen, 1964, 1967). Although freeze-drying of cultures offers many potential advantages, it is not yet applicable in practice to eukaryotic algae. For practical details of lyophilization see Chapter 4, this volume, on freeze-drying of bacteria.

V. LIQUID NITROGEN STORAGE

The outstanding successes of cryopreservation have been in the medical and veterinary sciences. Unfortunately, direct application of methods developed empirically for mammalian cell types are often unsatisfactory for the recovery of other organisms. This appears to be especially true for Protista. The approach taken in this laboratory is to develop simple practical methods of storage under liquid nitrogen which consistently yield high recoveries (>60%) upon thawing. Storage of Protista at high sub-zero temperatures has been reported (Schwarze, 1975; Takano *et al.*, 1973; Whitton, 1962). However, at these temperatures biochemical and biophysical processes may still occur, resulting in reduction in cellular viability with an increasing time of storage. Below $-139°C$ no growth of ice crystals occurs and the rate of other biophysical processes are too slow to affect cell survival.

For the long-term storage of cellular material it is essential to minimize the loss of viability, for the following reasons:

(1) practical difficulties may arise in re-establishing growth if the viable cell density falls below a minimum inoculation level;
(2) although the processes of cooling to and thawing from $-196°C$ are not considered to be mutagenic (Ashwood-Smith and Grant, 1977), the possibility exists that with low levels of recovery, pre-existing freezing-resistant mutants may be selected;

(3) cryopreservation can select subpopulations of a cell type (Knight *et al.*, 1972). This selectivity may have practical advantages in eliminating vegetative cells from resting stages (Leef and Gaertner, 1979; Peters and Sypherd, 1978), but must be avoided for purposes of genetic conservation.

Methods which have been reported to be successful for the storage and recovery of some freshwater and marine algae and for non-parasitic protozoa are listed in Tables II–IV. With some groups, such as Chlorococcales, methods exist by which it is relatively simple to recover a high proportion of viable cells upon thawing from liquid nitrogen. However, if these methods are applied directly to other groups of protists very poor results are obtained. In this section, some of the methods for assessing viability are first described; secondly, the factors which determine cellular viability following freezing and thawing are discussed, as this may allow the development of successful cryopreservation protocols for groups which have not previously been examined in detail. Finally, a standard method of cryopreservation devised at the Culture Centre of Algae and Protozoa, Cambridge will be described in detail to illustrate how it is often necessary to integrate several experimental approaches to obtain satisfactory recovery.

A. Assays of Viability

To assess cellular viability following any preservation method, it is important to measure some indicator of cell division, such as colony formation in or on agar, increase in the protein or chlorophyll content of a culture, or estimation of numbers by most probable number techniques. Indirect methods of estimating viability such as dye exclusion, oxygen evolution, uptake of vital stains and motility are available. Although these methods are simple, they usually overestimate the recovery potential; for instance, with *Tetrahymena pyriformis* a great many cells which are motile immediately upon thawing do not retain the capacity to divide (Osborne and Lee, 1975). Regrowth and division are definitive indicators of viability and whenever practical should be used.

1. Colony formation in agar
Many Protista will form discrete colonies on agar. However, with motile cells it is essential to assay viability within the agar; this latter technique will be described in detail. Logarithmic dilutions of the cells are prepared either in their growth medium or a balanced salts

Table II: Published methods for the cryopreservation of freshwater algae at −196°C.

Species	Growth Conditions before Freezing	Cryoprotective Additives	Cooling	Warming	Recovery	Reference
Cyanophyta (filamentous)	Desert soil samples	—	c.50°C min^{-1}	c.6°C min^{-1}	+	Cameron and Blank (1956)
Cyanophyta (7 strains from the Antarctic)	24°C, liquid culture	—	"Rapid"	"Slow"	+	Holm-Hansen (1963)
Chlamydomonas pseudagloe	4–8 d cultures washed from agar slopes	glycerol (10% v/v)	1°C min^{-1}	c.400°C min^{-1}	+	Hwang and Horneland (1965)
Chlamydomonas reinhardii	zygotes	—	N.S.	N.S.	>50%	Bennoun (1972)
C. reinhardii (3 mutant strains)	Mid-log phase cells precultured for 18 h in Me$_2$SO (1% v/v)	Me$_2$SO(5% v/v)	Two-step −20°C/120 min	"Rapid"	<0.01–5%	Gresshoff (1977)
C. reinhardii (11 mutant strains)	7–10 d cultures washed from agar slopes	Me$_2$SO(10% v/v)	1°C min^{-1}	"Rapid"	0.1%	Hwang and Huddock (1971)
C. reinhardii (10 mutant strains)	7–10 d cultures washed from agar slopes	Me$_2$SO(5% v/v)	1°C min^{-1}	"Rapid"	0.003–10%	McGrath and Daggett (1977)
C. reinhardii	7 d liquid culture at 20°C	Methanol (2.5M)	Two-step −25°C/60 min	90°C min^{-1}	60%	Morris et al. (1979)
Chlorella ellipsoidea	Synchronous culture cold hardened in the L$_2$ phase for 48 h at 3°C	—	Two-step −20°C/20 h	"Rapid"	>50%	Hatano et al. (1976a)

Table II—continued

Species	Growth Conditions before Freezing	Cryoprotective Additives	Cooling	Warming	Recovery	Reference
C. emersonii	21 d liquid culture at 4°C	—	212°C min^{-1}	90°C min^{-1}	40%	Morris (1976b)
C. emersonii	7 d liquid culture at 20°C	Methanol (2.5M)	10°C min^{-1}	90°C min^{-1}	80%	Morris *et al.* (1980)
C. protothecoides	7 d liquid culture at 20°C	—	0.3–200°C min^{-1}	90°C min^{-1}	98%	Morris (1976a)
Chlorococcales (7 strains from the Antarctic)	24°C liquid culture	—	"Rapid"	"Slow"	+	Holm-Hansen (1963)
Chlorococcales (13 strains)	4–8 d cultures washed from agar slopes	Glycerol (10% v/v)	1°C min^{-1}	c.400°C min^{-1}	+	Hwang and Horneland (1965)
Chlorococcales (300 strains)	35 d liquid culture at 20°C	Me$_2$SO (5% v/v)	Two-step −25°C/30 min	90°C min^{-1}	>50%	Morris (1978)
Euglena gracilis	4–8 d cultures washed from agar slopes	Glycerol (10% v/v)	1°C min^{-1}	c.400°C min^{-1}	+	Hwang and Horneland (1965)
E. gracilis (26 strains)	7 d liquid culture at 20°C	Methanol (2.5M)	0.3°C min^{-1}	90°C min^{-1}	>30%	Morris and Canning (1978)
Prototheca (8 strains)	14 d liquid culture at 4°C	—	0.3°C min^{-1}	90°C min^{-1}	>30%	Morris (1976c)
Spirulina platensis	Exponential phase cells under autotrophic conditions	Amino acids (1%) gum arabic (2–10%) gelatin (2–10%)	20–50°C min^{-1}	20–40°C min^{-1}	+	Takano *et al.* (1973)

N.S. = not stated.
+ = viable cells recorded but not quantified.

Table III: Published methods for the cryopreservation of marine algae at −196°C.

Species	Growth Conditions before Freezing	Cryoprotective Additive	Cooling	Warming	Recovery (%)	Reference
Chlorococcales (8 strains)	28 d liquid culture at 4°C	Me₂SO(5% v/v)	Two-step −30°C/15 min	"Rapid"	13–100	Ben-Amotz and Gilboa (1980a, b)
Cylindrotheca closterium	Log-phase culture at 25°C concentrated by centrifugation	Glycerol (5%) in half salinity medium	1°C min⁻¹	"Rapid"	100	Saks (1978)
Dunaliella quartolecta	Log-phase culture at 25°C concentrated by centrifugation	Glycerol (5%) in half salinity medium	1°C min⁻¹	"Rapid"	100	Saks (1978)
Nannochloris adamsii	Log-phase culture at 25°C concentrated by centrifugation	Glycerol (5%) in half salinity medium	1°C min⁻¹	"Rapid"	100	Saks (1978)
Nitzchia acicularis	Log-phase culture at 25°C concentrated by centrifugation	Glycerol (5%) in half salinity medium	1°C min⁻¹	"Rapid"	100	Saks (1978)
Phaeodactylum tricornatum	Log-phase culture at 25°C concentrated by centrifugation	Glycerol (5%) in half salinity medium	1°C min⁻¹	"Rapid"	95	Saks (1978)
P. tricornatum	28 d liquid culture at 4°C	Me₂SO (5% v/v)	Two-step −30°C/15 min	"Rapid"	36	Ben-Amotz and Gilboa (1980, a, b)
Platymonas succica	28 d liquid culture at 4°C	Me₂SO (5% v/v)	Two-step −30°C/15 min	"Rapid"	24	Ben-Amotz and Gilboa (1980, a, b)

Table IV: Published methods for the cryopreservation of non-parasitic protozoa at −196°C

Species	Growth Conditions before Freezing	Cryoprotective Additive	Cooling	Warming	Recovery (%)	Reference
Cyrtilophosis mucicola	Cysts	N.S.	N.S.	N.S.	+	Simon and Schneller (1973)
Naegleria sp.	Log-phase liquid culture at 25°C, concentrated by centrifugation	Me$_2$SO (5%)	1.3°C min^{-1}	"Rapid"	48–63	Simione and Daggett (1976)
Paramecium aurelia (150 stocks)	Liquid culture at 22–26°C, concentrated by centrifugation	Me$_2$SO (7.5%)	2–3°C min^{-1}	"Rapid"	1.3–16	Simon and Schneller (1973)
Tetrahymena pyriformis	Liquid culture, concentrated by centrifugation	Me$_2$SO (10%)	2–3°C min^{-1}	"Rapid" (35–39°C water bath better than 20–24°C)	+	Simon and Schneller (1973)
T. pyriformis	Young (2 d), liquid culture at 20°C	Me$_2$SO (1.4M)	1°C min^{-1} to −53°C → −196°C	"Rapid"	0.35	Osborne and Lee (1975)

N.S. = not stated.
+ = viable cells recorded but not quantified.

solution. One millilitre of the appropriate suspension is pipetted into a sterile petri dish, after which an appropriate nutrient solution containing agar (*c.* 1.5% w/v) at 40–42°C is poured in. The cells are dispersed through the agar by gentle agitation and, when set, the plate is inverted and kept under suitable cultural conditions. For cell types requiring long-term culture before the development of visible colonies, dehydration of the agar may be reduced by sealing the petri dish with 'Clingfilm'. For the assay of non-motile protista on the surface of agar the technique is similar to that described above, except that the diluted cell suspension is applied to the surface of a prepoured, solidified agar plate from which surface condensation has been removed. The cell suspension is distributed over the agar by using either the conventional spread plate method or the Miles and Misra technique (1958).

In the agar system, a colony can arise from either a single cell or from a cell clump. As many unicellular cells in liquid culture tend to form aggregates, it is essential to estimate cellular multiplicity in order to determine accurately the total cell number. The frequency of cellular aggregation is determined microscopically from three samples, at least 500 cells being counted for each treatment. The multiplicity of the sample is then defined as:

$$\text{Multiplicity} = \frac{\text{Number of cells}}{\text{Number of groups}}$$

The total cell number is then:

Number of colonies × Multiplicity

The percentage of viability is:

$$\frac{\text{Total number of cells before treatment}}{\text{Total number of cells after treatment}} \times 100$$

Following freezing and thawing there is often a breakdown of cell aggregates; typical counts are illustrated in Fig. 1. In this example there is a reduction in multiplicity from 1.21 to 1.09. The recovery following freezing and thawing is 86% if cellular aggregation is taken into consideration or 95% if it is ignored. The extent of the reduction in multiplicity is a factor of cell type, rates of cooling and warming, concentration and type of additive; therefore, whenever accuracy is required the multiplicity should be determined for each frozen and thawed sample. If this is not taken into consideration, results should be expressed as colony forming units.

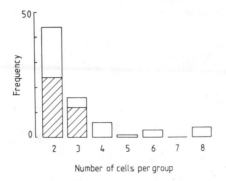

Fig. 1: Distribution of cellular aggregation in a 7-day culture of *Chlorella prototothecoides*. Open figures unfrozen, hatched figures following freezing and thawing. The number of single cells was 480 unfrozen, 505 frozen and thawed.

2. Most probable number techniques

With organisms which will not readily form colonies in or on agar, viable cell numbers may be determined by most probable number techniques in liquid media. For details of these methods as originally developed using tubes of media, see Swaroop (1938). To economize, methods using microtitre plates have been developed for the assay of ciliates (Heaf and Lee, 1971) and may be successfully applied to other protists.

B. *Factors Affecting Viability Following Cryopreservation*

The many variables which determine the response of Protista to the stresses of freezing and thawing can be separated into two classes:

(1) Intrinsic or cellular factors including the choice of cellular material, growth temperatures, age of culture, growth medium and post-thaw culture conditions.
(2) Extrinsic or physical determinants, such as type and concentration of cryoprotective compound, rates of cooling and warming and the final temperature attained.

Although many of these factors are interrelated and all contribute to the viability of cells upon thawing, they will be discussed separately.

1. Intrinsic factors

Cellular viability following freezing is partly determined by the choice of cellular material and the culture conditions before and after freezing. Natural resting stages, with a low water content, are more resistant to injury than are hydrated vegetative forms. For example, zygotes of *Chlamydomonas reinhardii* are resistant to freezing to and thawing from −196°C (Bennoun, 1972), in contrast to the vegetative cells which are extremely susceptible to freezing injury (Morris *et al.*, 1981). If possible, therefore, a dehydrated form such as a zygote or cyst should be selected for preservation. However, this often cannot be done and it is important to determine the growth conditions which produce cells in a state most resistant to freezing injury. The culture conditions which have been most frequently examined include those below.

(a) Growth temperature. The freezing resistance of some freshwater (Hatano *et al.*, 1976a, b; 1978; Morris, 1976b, c) and marine (Ben-Amotz and Gilboa, 1980b) unicellular algae and ciliated protozoa (Polyansky, 1963) increases with a decrease in ambient temperature. This phenomenon is analogous to the cold-hardening process which occurs naturally in higher plants in autumn, allowing

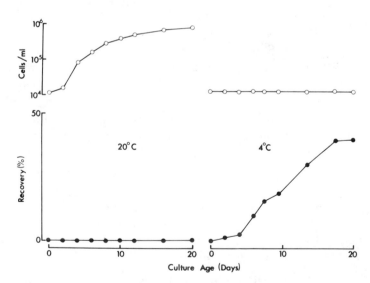

Fig. 2: Recovery (%) of *Chlorella emersonii* following freezing to and thawing from −196°C on successive days of culture at 20 and 4°C.

survival through the extremes of winter.

A typical pattern of cold acclimatization is observed with the unicellular green alga *Chlorella emersonii* (Fig. 2). Although at 4°C there is no net increase in cell number, the biochemical adaptations which allow metabolism to continue at this reduced temperature increases the resistance to freezing injury compared with cells cultured at 20°C.

Cold acclimatization is thus potentially important in increasing the degree of recovery following freezing and thawing. Three types of response may be observed with Protista:

(1) Following a period of cold acclimatization the cells become resistant to freezing injury (Hatano *et al.*, 1976a, b; 1978; Morris 1976b, c; Ben-Amotz and Gilboa, 1980b).

(2) Cells adapt metabolically to a reduced temperature with no significant increase in freezing tolerance (*Euglena*, *Chlamydomonas*).

(3) Cellular injury may occur during prolonged incubation below a critical temperature, a phenomenon analogous to chilling injury in higher plants (Levitt, 1980).

Practically, it has been found preferable to transfer an early stationary phase culture to the hardening temperature rather than to inoculate cells into cold media. This results in a denser cell suspension for subsequent cryopreservation.

(b) Age of culture. When suspension cultures are grown by batch methods they pass through definable phases of culture. With cultures of algae, cells from the actively dividing exponential phase are more sensitive to freezing injury than are cells from the older stationary phase (Morris, 1978). In the late stationary phase of culture, cell growth is limited by many nutritional and physical factors; under these conditions cells accumulate lipid, become less vacuolated and modifications to the membrane fatty acids occur (Morris and Clarke, 1978).

(c) Nutrient limitation. Reduction of the growth rate of algal cultures by the limitation of nutrients has been demonstrated to increase freezing tolerance. Depletion of nitrate (Morris and Clarke, 1978) and bicarbonate (Ben-Amotz and Gilboa, 1980b) induce the highest degree of resistance.

As a general principle, the susceptibility of algal cells to freezing injury is associated with the presence of a large vacuole. The degree of vacuolation is reduced following the accumulation of storage lipid at low growth rates. Freezing tolerance is thus increased following treatments which reduce the growth rate, such as limitation of nutrients, reduction in temperature, addition of metabolic inhibitors, or use of hypertonic growth media.

2. Extrinsic factors

When the maximum intrinsic tolerance has been achieved by modifications in the growth conditions, the cellular survival is then determined by the complex interactions between the physical determinants of the freezing and thawing processes and the addition, if any, of protective compounds.

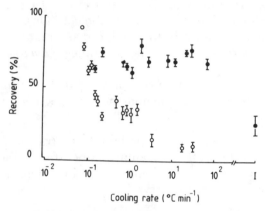

Fig. 3: Recovery (%) following undercooling (i.e. in the absence of freezing and thawing) at different rates to −10°C. *Amoeba* spp. strain Borok (○) and strain Warsaw (●).

(a) Cold-shock. Many cell types are damaged by a rapid reduction in temperature; this has been termed cold-shock. Protista known to be susceptible include *Amoeba* (Fig. 3), *Tetrahymena* (Fuller *et al.*, 1982), *Chlamydomonas* (Morris *et al.*, 1983), *Blepharisma* (Giese, 1973) and *Anacystis* (Siva *et al.*, 1977). The first step in any cryopreservation method is to cool the sample to a temperature at which ice nucleation is initiated and cells are thus exposed to the stress of temperature reduction. With cell types that are sensitive to this stress, injury can be avoided by:

(1) Reducing the rate of temperature reduction, as cold-shock stress is minimized at slow rates of cooling, although the absolute values of "slow" and "rapid" cooling vary with the cell type (Fig. 3).

(2)   Some cell types are sensitive to cold-shock only when under-going rapid division (Morris, 1976a). Survival is thus increased by using stationary phase cultures.

(3)   Injury is a function of the growth temperature, suspensions of *Anacystis nidulans* being extremely susceptible to cold-shock when cultured at 35°C but resistant when grown at 25°C (Siva *et al.*, 1977).

(b) Rates of cooling. Some cells are intrinsically resistant to injury following freezing to and thawing from −196°C, the survival usually being dependent on the rate of cooling (Fig. 4). With *Scenedesmus quadricauda* and *Chlamydomonas nivalis* optimal rates of cooling of 2 and 10°C min^{-1} were observed, with survival decreasing at both faster and slower rates of cooling.

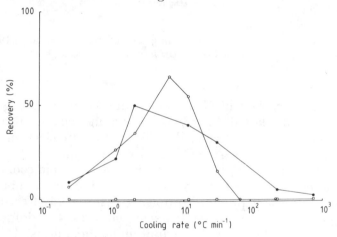

Fig. 4: Recovery (%) of algal cells after cooling at different rates to −196°C. *Chlamydomonas nivalis* (○), *Scenedesmus quaricauda* (●), and *Chlorella emersonii* (□).

However, with the majority of algal cells the survival upon thawing from −196°C is very low and is significantly increased only by the addition of protective compounds or following pretreatments such as cold-hardening. With *Chlorella emersonii* cold acclimatization increased the recovery at rapid rates of cooling. In contrast, the recovery of the apoplastidic alga *Prototheca*, following a similar period of cold acclimatization, was maximal at the slowest rate of cooling examined (Fig. 5). Therefore, in any comparison of the effect of growth treatments upon cellular freezing tolerance a range of cooling rates must be examined, as protective effects may not be observed at any single cooling rate.

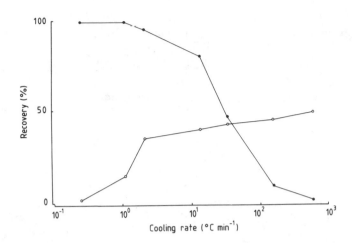

Fig. 5: Recovery (%) following cooling at different rates to −196°C of cold-hardened *Chlorella emersonii* (○) and *Prototheca chlorelloides* (●).

Ultra-rapid cooling of algal cells, achieved by spraying a cell suspension into liquid nitrogen, results in cells in which no intracellular ice can be detected by freeze-fracture electron microscopy and, provided warming is also rapid, complete recovery of viability (Plattner *et al.*, 1972). The advantage of rapid cooling as a preservation method is that cryoprotective additives are not essential. However, many practical problems exist in spray-freezing and retrieving adequate amounts of axenic cultures. In addition, the technique is applicable only to unicellular organisms. In larger specimens, the rate of heat transfer is slower and intracellular ice nucleation and crystal growth are inevitable.

(c) Cryoprotective additives. For most cell types it is necessary to include compounds, so-called cryoprotective additives, to reduce the injury upon thawing. The effects of a large number of potential cryoprotective compounds on the recovery of frozen and thawed mammalian cells, especially erythrocytes and spermatozoa, have been examined on an empirical basis. Unfortunately, few such studies have been reported for other cell types and it has been generally assumed that the additives glycerol or dimethyl sulphoxide (Me$_2$SO) would be optimally effective. However, in the case of Protista it is now apparent that other compounds may be more suitable as cryoprotectants (Table V). With other algae, the amino

Table V: Median lethal temperatures (LT$_{50}$)* for *Euglena gracilis* following freezing at a rate of 0.25°C min^{-1} in different additives.

Additive	Concentration	LT$_{50}$(°C)
None	—	−5.3
Dimethyl sulphoxide	10% (v/v)	−10.2
Ethanol	10% (v/v)	−15
Methanol	10% (v/v)	>−50
Polyvinylpyrrolidone	10% (w/v)	−4.1
Sucrose	7.5% (w/v)	−4.0
Glycerol	5% (w/v)	−3.4
Glucose	5% (w/v)	−2.8

* The median lethal temperature was defined as that at which 50% of the cells were killed during freezing and thawing under standard cooling and warming conditions.

acid proline has been demonstrated to be a very effective cryo-protective additive (Meyer, unpublished observations).

The morphological response of *Euglena gracilis* during exposure to different potentially cryoprotective additives has been described previously (Morris, 1980). For *E. gracilis* the non-penetrating, low molecular weight additives (glucose, glycerol, and sucrose) were more damaging than were penetrating (ethanol, methanol, and dimethyl sulphoxide) or higher molecular weight (polyvinyl-pyrrolidone) additives. The concentrations found to be relatively non-toxic (>75% recovery following a 15 min exposure at 20°C) are shown in Table V.

At a rate of cooling of 0.25°C min^{-1} the extracellular additives increased cellular injury, whilst the penetrating additives were protective. The most effective cryoprotectant, at this rate of cooling, was methanol with a median lethal temperature (LT$_{50}$) below −50°C. The protection afforded by additives during freezing and thawing is dependent on the rate of cooling. With *E. gracilis* cooled in methanol (2.5 M) an optimal rate of cooling of 0.34°C min^{-1} was observed, with cell recovery falling at both higher and lower rates (Fig. 6).

(d) Two-step cooling. In this technique, which is best considered as interrupted rapid cooling, freezing occurs during an initial period of rapid cooling to a constant holding temperature. After maintenance at this temperature the sample is then cooled rapidly into liquid nitrogen (McGann and Farrant, 1976). This method has several practical advantages: it is simple to carry out, requires no specialized

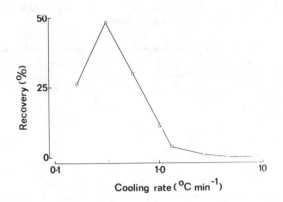

Fig. 6: Recovery (%) of *Euglena gracilis* suspended in methanol (2.5 M) after cooling at different rates to −196°C.

controlled cooling-rate equipment and low concentrations of additive are effective.

The recovery following two-step cooling of *E. gracilis* is shown in Fig. 7: the cells were initially suspended in methanol and frozen rapidly to −30°C before either thawing directly or plunging into liquid nitrogen and then thawing. The survival of cells thawed

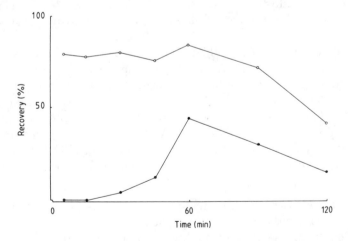

Fig. 7: Recovery (%) of *Euglena gracilis* frozen in methanol (2.5 M) at −30°C for different times before either thawing (○) or plunging to −196°C before thawing (●).

directly from −30°C decreased steadily with increasing time of exposure. The recovery from −196°C initially increased with time at −30°C, then slowly decreased in a manner parallel to the loss of viability at the holding temperature. Survival from −196°C is not affected by adoption of a convenient rate of cooling produced by immersing the sample in a constant temperature bath, since protection is acquired with time at the holding temperature once that temperature has been reached.

(e) Warming Rate. In all studies where the effects of rate of warming from −196°C on the survival of Protista have been examined, rapid rates gave maximal survival.

C. Cryopreservation at CCAP

A flow diagram of a method developed for the cryopreservation of certain algal cultures at CCAP is outlined in Fig. 8. Several

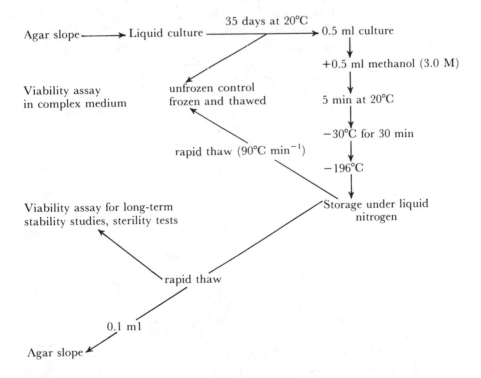

Fig. 8: Flow diagram of cryopreservation at the Culture Centre of Algae and Protozoa, Cambridge.

experimental approaches are integrated into a practical method which is both simple to carry out and yields consistently high recoveries upon thawing.

Cells from an agar slope are transferred to liquid medium and cultured into the late stationary phase (35 days at 20°C), at which stage of culture the cells are non-vacuolated and are most intrinsically resistant to freezing injury. Alternatively, the cellular freezing tolerance can be increased following cold acclimatization or by the limitation of nutrients; however, following these last treatments, the cultures have a low cell density and it is thus more difficult to re-establish growth upon thawing than with senescent cultures of higher cell density.

Cells from liquid culture are used without further preparation. 0.5 ml of the cell suspension is added to 0.5 ml of methanol (3.0 M, prepared in growth medium) in a 12 × 35 mm sterile polypropylene tube (Gibco-Europe Ltd (Nunc)). Methanol cannot be autoclaved and, if required sterile, must be filtered. However, methanol dissolves some filters, such as Millipore (Millipore (UK) Ltd) and therefore a membrane resistant to methanol, such as Nuclepore (Nuclepore Corp.) must be used. The cells are exposed to the 1.5M methanol solution for 5 min at 20°C to ensure complete uptake of the additive. The ampoules are then frozen by a two-step method, first being placed in an alcohol bath (Camlab Ltd) maintained at −30°C for 15 min, and then, being transferred directly into liquid nitrogen. If a specialized low temperature bath is not available, coolant can be placed in a deep-freezer, provided the volume of the coolant is large enough for any temperature fluctuation to be minimized. Thawing is by rapid agitation of the ampoule in a water bath at 25°C until the last visible crystal of ice has disappeared; viability is then determined.

Long-term storage of ampoules is under liquid nitrogen, 12 ampoules of each strain being routinely preserved. At CCAP, a recovery of 60% is the minimum accepted for long-term maintenance. Using the method outlined, over 400 strains of algae are now successfully stored under liquid nitrogen. It must be emphasized that this technique, although satisfactory with many strains of Chlorococcales and Euglenophyceae, is less successful with other groups of Protista. Improvement with other cell types can best be achieved by an understanding of the basic principles of cryobiology rather than by following recipes.

VI. CULTURE MEDIA FOR ALGAE AND PROTOZOA

A. Actinosphaerium *Medium*

Stock solution: 500 ml glass-distilled water is heated to 30°C and 0.145 g Na EDTA (disodium salt) is added. The solution is allowed to cool.

Final medium: to make 1 litre, river water is collected, filtered immediately, and then autoclaved. In pouring autoclaved river water, any precipitate which has formed is discarded. One ml stock solution is added to 999 ml filtered, autoclaved river water.

Before inoculating new cultures, the medium is aerated immediately before use by bubbling air through it for at least 2 h.

B. Amoeba Saline

Stock solutions: each of the following is dissolved separately in glass-distilled water made up to 100 ml:

NaCl	1.20 g	Na_2HPO_4	1.42 g
$MgSO_4.7H_2O$	0.04 g	KH_2PO_4	1.36 g
$CaCl_2.2H_2O$	0.04 g		

Final medium: 10 ml of each stock solution is made up to 1 litre with glass-distilled water (Page, 1967).

C. Biphasic Media

These simple media promote good growth of many algae and protozoa.

Basic medium: a layer about 1 cm deep of good calcareous garden soil is put into a test tube or jar. (The use of mud from ponds or rivers is seldom satisfactory.) Water is added to a depth of 7–10 cm and the container plugged or covered and steamed for 1 h (longer for larger vessels) on each of 2 consecutive days. The medium is allowed to stand for a further day before inoculating, when the pH should be between 7 and 8.

Many variations of this basic medium are possible. The garden soil can be replaced by calcareous clay. The addition beneath the soil of a little (about 3% of the volume of soil) calcium carbonate or ammonium magnesium phosphate is recommended—the former for many eutrophic chlorophyceae, the latter for many green euglenoids.

Sphagnum peat may be added or may replace the soil when growing forms from acid habitats. The addition of starch below the soil stimulates growth of *Polytoma* and *Astasia* species. A grain of pearl barley, rice, or wheat placed on the surface of the medium produces a bacterial flora forming suitable food for many ciliates.

When selecting soils it is advisable to take a fair-sized sample (about 0.04 m^3) and to pass it through a sieve of about 1 cm mesh.

D. Bold's Basal Medium

This medium supports growth of a wide range of algae and may be supplemented with 5% soil extract.

Stock solutions: six stock solutions (i–vi) are prepared, each containing one of the following salts dissolved in 400 ml of distilled water:

(i)	$NaNO_3$	10.0 g	(iv)	K_2HPO_4	3.0 g
(ii)	$CaCl_2.2H_2O$	1.0 g	(v)	KH_2PO_4	7.0 g
(iii)	$MgSO_4.7H_2O$	3.0 g	(vi)	NaCl	1.0 g

Four trace element solutions (vii–x) are prepared by dissolving the following in 1 litre of distilled water.

(vii)	EDTA	50 g
	KOH	31 g
(viii)	$FeSO_4.7H_2O$	4.98 g
	(use acidified H_2O: 1 ml of H_2SO_4 added to 999 ml distilled water)	
(ix)	H_3BO_3	11.42 g
(x)	$ZnSO.7H_2O$	8.82 g
	$MnCl_2.4H_2O$	1.44 g
	MoO_3	0.71 g
	$CuSO_4.5H_2O$	1.57 g
	$Co(NO_3)_2$	0.49 g

Final medium: 10 ml of each stock solution (i) to (vi) and 1 ml of each stock solution (vii) to (x) are added to 936 ml of distilled water (Bischoff and Bold, 1963).

E. CP Fluid

0.1 g Cerophyl
P & J solution (see medium H) to make 1 litre
The Cerophyl is added to P & J solution and boiled for 5 min. The particulate matter is filtered out and the volume restored with glass-distilled water.

F. Erdschreiber Medium

Many marine algae will grow in this medium. It is also suitable for some marine ciliates and amoebae.

Natural sea water	1 litre
(or artificial sea salts)	
Soil extract solution	50.0 ml
$NaNO_3$	0.2 g
$Na_2HPO_4.12H_2O$	0.03 g

G. Euglena gracilis *Medium*

This medium is also suitable for axenic algal cultures, particularly those requiring acetate.

	g/100 ml glass-distilled water
Sodium acetate (hydrated)	0.1
Beef extract (Lab-lemco powder)	0.1
Yeast extract	0.2
Bacto tryptone	0.2
$CaCl_2$	0.001
Agar (optional)	1.0

H. Prescott and James' (P & J)

Stock solutions:

		g/100 ml glass-distilled water
(i)	$CaCl_2.2H_2O$	0.433
	KCl	0.162
(ii)	K_2HPO_4	0.512
(iii)	$MgSO_4.7H_2O$	0.280

Final medium: 1 ml of each of (i), (ii), and (iii) is added to 997 ml glass-distilled water (Prescott and James, 1955).

I. Non-nutrient Agar

15 g of Oxoid agar No. 1 is added to 1 litre of Amoeba saline (medium B) and stirred until dissolved. The medium is then autoclaved and dispensed into petri dishes.

J. Proteose Peptone

This medium is suitable for bacteria-free cultures of algae.

	g/100 ml glass-distilled water
Proteose peptone (Difco)	0.1
KNO_3	0.02
K_2HPO_4	0.002
$MgSO_4.7H_2O$	0.002
Agar (optional)	1.0

K. Soil Extract Medium

This medium is suitable for many algae, particularly those cultures in which bacteria are present.

	g/100 ml water
KNO_3	0.02
K_2HPO_4	0.002
$MgSO_4.7H_2O$	0.002
Agar (optional)	1.0

Soil extract stock solution 10% by volume.

The soil extract stock solution is made by heating for 2 h in a steamer a calcareous garden loam with twice its volume of supernatant water, or by autoclaving for 15 min. As repeated autoclaving is deleterious, it is recommended that a number of small containers of stock solution be prepared, each of a size appropriate to making a batch of final medium.

L. Kratz and Myers Medium C

Many blue-green algae will grow well on this medium.

	g/litre
$MgSO_4.7H_2O$	0.25
K_2HPO_4	1.00
$Ca(NO_3)_2.4H_2O$	0.025
KNO_3	1.000
$Na_3citrate.2H_2O$	0.165
$Fe(SO_4)_3.6H_2O$	0.004
Trace element stock solution*	1.0 ml
Agar (Optional)	10.00 g

* Trace element solution:

	g/litre
H_3BO_3	2.86
$MgCl_2.4H_2O$	1.81
$ZnSO_4.7H_2O$	0.222
$MoO_3(85\%)$	0.0177
$CuSO_4.5H_2O$	0.079

(Kratz and Myers, 1955)

14

Maintenance of Parasitic Protozoa by Cryopreservation

E. R. JAMES

London School of Hygiene and Tropical Medicine
Winches Farm Field Station
St. Albans, UK

I. INTRODUCTION

It was recognized long ago (Coggeshall, 1939) that the techniques in existence for laboratory maintenance of parasitic protozoa were frequently tedious and expensive, and that irreversible biological changes often occurred in the material being maintained. Cryopreservation overcomes these problems and, although the development of techniques for cryopreservation of parasitic protozoa has not advanced as fast as the science of cryobiology, cryopreservation has now become an indispensible tool in this field; nevertheless, many of the techniques described are still obviously suboptimal.

The techniques described for cryopreservation of parasitic protozoa are many and various, reflecting both the different requirements of the different species of protozoa as well as the funds

MAINTENANCE OF MICROORGANISMS
ISBN 0 12 410350 2

available to the various workers and the local conditions in which they have to operate. It cannot be stressed strongly enough that a set of conditions that leads to good survival of one species is very likely to be wholly inappropriate for another species and a particular cryopreservation protocol should only be treated as a basic guideline.

Survival of cryopreserved parasitic protozoa has usually been extremely low, even in many of the more recent studies which have incorporated cryoprotectants and have defined the cooling and warming procedures. This has often not mattered too much since the parasites have been capable of rapidly reproducing in culture or in their host following injection. However, where survival levels are particularly low this will impose a selective pressure on the parasite population. It is imperative, therefore, to use a technique which aims to give the highest possible survival rate of the particular parasite species.

During slow cooling, water in the suspending medium is converted into ice and the salts become concentrated. This creates an osmotic imbalance across the cell's outer membrane and so the cell shrinks. The sensitivity of different cell types to the concentrated levels of solutes produced during freezing and to the different cryoprotectant compounds varies considerably.

At rates of cooling which are too fast to allow the cell to equilibrate by shrinkage, intracellular freezing occurs. The size of the intracellular ice crystals which form depends on the cooling rate—the crystals being smaller when faster rates are used. At very fast rates the ice crystals may be so small that cell damage is avoided.

Very rapid cooling rates coupled with low storage temperatures and very fast warming rates prevent ice from forming altogether. It is almost impossible to achieve the very rapid cooling needed to vitrify pure water; however, for aqueous solutions the addition of cryoprotectants considerably slows the cooling rate required for vitrification, although relatively high concentrations (>30% v/v) of additive have to be used. Except for *Naegleria* (Luyet and Gehenio, 1954), very rapid cooling has not been used much with parasitic protozoa.

Between the extremes of slow and fast cooling there lies an optimum cooling rate for a particular type of cell. This optimum cooling rate is typically shifted to a slower rate as increasing concentrations of cryoprotectant are used. Some cells that are particularly small in size and/or have a surface membrane that is highly permeable to water will dehydrate sufficiently, even at

Table I: Synopsis of published successful techniques for cryopreservation of parasitic protozoa giving good levels of survival.

Parasite	Container and Vol.	Additive Type and Conc.	Equilibration Time and Temp.	Cooling Rate	Thawing Rate	Dilution Procedure	Reference
Babesia bigemina	Plastic straw 0.5 ml	Me_2SO 14.2% v/v	30 min, 0°C	82°C min^{-1} to −60°C, fast to −196°C	Fast, in water bath at 40°C	None	Dalgliesh and Mellors (1974)
Babesia bovis (culture & extracellular forms)	— 1.0 ml	PVP^a 10% v/v	—	20°C min^{-1} to −70°C, slow to −196°C	Fast, in water bath at 37°C	1:50 in Pucks Saline G	Palmer et al. (1982)
Babesia rodhaini	Glass capillary 25 μl	Me_2SO 10.7% v/v	0°C	265°C min^{-1}	Fast, in water bath at 40°C	None	Dalgliesh et al. (1976)
Eimeria (sporozoites)	glass ampoule 1 ml	glycerol 7.5% v/v	20 min, 20°C	1°C min^{-1} to −70°C, fast to −196°C	Fast, in water bath at 37°C	None	Norton et al. (1968)
Eimeria (sporocysts)	glass ampoule 1 ml	Me_2SO 7.5% v/v	15 min, 20°C	1°C min^{-1} to −70°C, fast to −196°C	Fast, in water bath at 37°C	None	Norton and Joyner (1968)
Entamoeba histolytica (trophozoites)	glass ampoule 1 ml	Me_2SO 7.5% v/v	15 min, 37°C	1°C min^{-1} to −60°C, fast to −196°C	Fast, in water bath at 37°C	None	Farri (1978)
Entamoeba histolytica (trophozoites)	glass ampoule 1 ml	ME_2SO 7.5% v/v	15 min, 37°C	Fast to −25°C, hold 20 min, fast to −196°C	Fast, in water bath at 37°C	None	Farri (1978)
Giardia intestinalis	plastic ampoule 0.25 ml	Me_2SO 7.5% v/v	15 min, 23°C	1°C min^{-1} to −50°C, fast to −196°C	Fast, in water bath at 37°C	1:50 in Diamond's medium	Warhurst and Wright (1979)

Table I—continued

Parasite	Container and Vol.	Additive Type and Conc.	Equilibration Time and Temp.	Cooling Rate	Thawing Rate	Dilution Procedure	Reference
Leishmania tropica var major	corked glass tube 0.5 ml	Me$_2$SO 10.65% v/v	— 0°C	1.9°C min^{-1} to −65°C, fast to −196°C	600–800°C min^{-1} in water bath at 37°C	>1:10 in 10 mM glucose saline	Callow and Farrant (1973)
Leishmania spp.	glass capillary 25 μl	glycerol 7.5% v/v	10 min, 20°C	0.7°C min^{-1} to −60°C, rapid to −196°C	Fast, in air at 20°C	None	Lumsden et al. (1973)
Naegleria	glass ampoule 0.5 ml	Me$_2$SO 5.0% v/v	30 min, 20°C	1.3°C min^{-1} to −55°C, fast to −196°C	Fast, in water bath at 35°C	—	Simione and Daggett (1976)
Plasmodium (trophozoites)							
P. chabaudi (rodent)	glass capillary 25 μl	glycerol 10% v/v	20 min, 20°C	Plunge into LN at 3,600°C min	Rapid, in water bath at 37°C	15% w/v glucose	Mutetwa (1983)
P. falciparum	glass capillary 25 μl	glycerol 10% v/v	20 min, 20°C	Cap. in glass tube at 300°C min^{-1} to −196°C	Rapid, in water bath at 37°C	15% w/v glucose	Mutetwa (1983)
P. galinaceum (avian)	glass capillary 25 μl	glycerol 10% v/v	20 min, 20°C	1°C min^{-1} to −80°C, rapid to −196°C	Rapid, in water bath at 37°C	15% w/v glucose	Mutetwa (1983)
Plasmodium sporozoites							
P. berghei	glass tube 1 ml	HES[b] 10% v/v serum 50% v/v	— 0°C	50°C min^{-1}	300°C min^{-1}	—	Leef et al. (1979)
Theileria parva	glass cap. 100 μl, or tube 2.5 ml	glycerol 7.5% v/v	—	slow (1°C min^{-1}?) to −80°C, fast to −196°C	Fast, in water bath at 37°C	None	Cunningham et al. (1963)

Table I—continued

Parasite	Container and Vol.	Additive Type and Conc.	Equilibration Time and Temp.	Cooling Rate	Thawing Rate	Dilution Procedure	Reference
Toxoplasma gondii	glass ampoule 2 ml	glycerol 5% v/v	— —	0.3°C min^{-1} to −79°C	Fast, in water bath at 37°C	None	Eyles et al. (1956)
Toxoplasma gondii	glass ampoule 1 ml	glycerol 5% v/v	— —	1°C min^{-1} to −160°C	Fast, in water bath at 37°C	1 in 5 in Minimum Essential Medium	Bollinger et al. (1974)
Trichomonas vaginalis	glass tube 1 ml	Me$_2$SO 7.5 v/v	0 min, 25°C	In −30°C freezer 90 min, then to −75°C freezer	Fast, in water bath at 37°C	—	Miyata (1975)
Trichomonas vaginalis	glass ampoule 2 ml	Me$_2$SO 5% v/v	20 min, 20°C	1°C min^{-1} to −35°C, fast to −196°C	Fast, in water bath at 45°C	1 in 20 in Diamond's medium	Ivey (1975)
Trypanosoma spp.	glass capillary 25 µl	glycerol 7.5% v/v	10 min, 20°C	0.7°C min^{-1} to −60°C, rapid to −196°C or −140°C	Fast, in air at 20°C	None	Lumsden et al. (1973)

[a] polyvinylpyrrolidone.   [b] hydroxyethyl starch.

cooling rates of several hundred or even several thousand degrees per minute. Mammalian red cells survive well at very fast cooling rates and, hence, the best cooling rates for many of the intraerythrocytic parasites tend to be in this range. However, many other parasitic protozoa appear to survive best with slower cooling rates in the region of 1°C min^{-1}. Table I gives a synopsis of those published techniques for the cryopreservation of parasitic protozoa which have attempted to define an optimum set of conditions.

Two-step cooling, in which cells undergo slow or rapid cooling to a particular intermediate temperature, and, after a holding period, further rapid cooling to the storage temperature, can reduce some of the potentially damaging effects. The cells become partially shrunken during the slow cooling step or at the intermediate holding temperature if the initial cooling step is rapid; then, during the second rapid cooling step to the storage temperature, the intracellular contents are either converted directly into a glass phase or, if ice crystals are formed, they may be too small to be damaging. The intermediate temperature is critical: at temperatures above the optimum the cells are not shrunken enough, so that large internal ice crystals form during the second cooling step. At temperatures below the optimum or with extensive periods at the intermediate temperature, the cells become too dehydrated and/or the concentration of solutes is damaging. For parasitic protozoa, two-step cooling was first applied to trypanosomes (Cunningham *et al.*, 1964). It has also been used successfully for *Plasmodium* (Wilson *et al.*, 1977) and for *Entamoeba* (Farri, 1978).

During final storage, deterioration of cells is generally more likely to occur at high temperatures, particularly above −136°C. At lower temperatures almost indefinite storage is possible, with little or no deterioration. Special care should be taken when transferring samples between storage vessels to protect them from warming by even a few degrees. Also, when removing individual samples for thawing, other samples in the same rack or canister should be prevented from warming.

Slow warming is usually damaging, but can sometimes be better than rapid warming if the cooling rate used was slow. However, optimum survival following rapid cooling is obtained by rapid warming—this indicates that at least some of the damage associated with intracellular ice formation occurs during warming. If small ice crystals do form during rapid cooling these will grow into larger damaging crystals during slow warming.

Few of the many techniques described for the cryopreservation of parasitic protozoa have been derived by careful empirical research. The trypanosomes have been most extensively studied. Following investigations by Cunningham, Lumsden and others a technique of slow cooling (between 0.7 and 5°C min^{-1} from 0 to -60°C) and fast warming (approximately between 8 and 2000°C min^{-1} from -60 to 0°C) was developed which gave survival levels that were little different from unfrozen controls when assayed by an infective dose test (ID_{50} ml^{-1}) (Webber *et al.*, 1961). The slow cooling technique described below is based on this method.

II. PREPARATION

Although different species of parasites require different cryopreservation techniques for optimal survival, the overriding consideration is probably the apparatus available for storage. Because the use of ampoules restricts the upper range of cooling rates which can be used to about 200–400°C min^{-1}, depending on the material from which the ampoule is constructed and the volume being preserved, many workers cryopreserve all their organisms in glass capillary tubes. However, there are certain advantages and disadvantages to using capillary tubes or ampoules.

Glass capillaries and ampoules are flame-sealed and thus they can be used for the human pathogenic protozoa and for material which is required to be sterile. The disadvantage of glass is that it can shatter, so there may be a risk of infection. Also, if improperly sealed, liquid nitrogen may enter the tube or ampoule during storage and this will almost certainly cause it to explode upon thawing. Some cryoprotectants, such as sucrose and certain polymers, have lower coefficients of expansion than glass and thus can cause glass ampoules, and particularly glass capillaries, to shatter during cooling. Polypropylene ampoules (e.g. Nunc cryotubes, Gibco-Europe Ltd (Nunc)) are favoured by some workers because they do not crack or fracture; however, although the silicone rubber rings are seated within the caps and the manufacturers say leakage cannot occur, even with the caps screwed down tightly they can occasionally leak and sterility is more difficult to maintain. Nevertheless, these tubes are safer because, if liquid nitrogen does penetrate during storage, they are very unlikely to explode during thawing.

The plastic straws used for storage of spermatozoa are usually not at all suitable since they frequently fracture or crack, or the metal or plastic plugs used to seal the ends drop out. Again, entry of liquid nitrogen into one of these straws is dangerous as the plastic plugs or ball bearings can be blown out of the ends at high velocity during thawing.

A. Capillary Tubes

The glass capillary tubes should be cleaned in chromic acid (3 parts saturated potassium dichromate solution:1 part sulphuric acid) (detergents do not produce adequate cleaning) and rinsed three times in tap water followed by three times in demineralized or double glass-distilled water prior to use. After drying in a warm oven they should be stored in test tubes (e.g. 125 × 16 mm) sealed with an aluminium foil cap. If sterilization of the tubes is required, it is suggested (Lumsden *et al.*, 1973) that autoclaving at 121°C be used, since capillaries sterilized by dry-heat do not fill so easily with the parasite suspensions. Final drying of autoclaved tubes can be carried out in a warm oven.

If a large number of tubes is to be frozen in one batch, it is convenient to have pivotted tube holders constructed accommodating 25 or more tubes each. Tubes of length 100 mm and external diameter 1mm are commonly used (light wall type, Plowden and Thompson Ltd), but the size is not critical. The suspension of parasites and cryoprotectant is drawn into a Pasteur pipette, which is used to fill about 20 mm of the length of each tube as rapidly as possible. The individual capillaries or the tube holder is then tilted so that the suspensions run to about the centre of the length of the tubes. Each tube is then sealed with a microburner (e.g. Soudagaz S, Baird and Tatlock (London) Ltd) starting with the end through which the suspension was introduced, this being more difficult to seal. The seal can then be checked if the tubes are tilted into a vertical position. Improper sealing can be identified by the suspension running down inside the tubes, and these tubes should be resealed or discarded. The other ends of the tubes are then sealed with the microburner, avoiding the formation of a terminal bubble or bead if possible. This can be achieved by pushing the tube 3–4 mm into the burner flame to heat the air in the end of the tube, and then by sealing the extreme tip. Perfect sealing of the tubes is important to conserve sterility and when storing the tubes in liquid nitrogen.

With experience, the approximate time taken to add glycerol to the suspension of parasites and to fill 25 tubes is in the region of 10 minutes; this is thus the minimum equilibration time in the cryoprotectant before commencing the cooling procedure.

B. Ampoules

Ampoules are often more convenient containers than capillaries for some applications and some parasites. Larger volumes can be frozen and stored more easily and, being larger than capillaries, they are also easier to label, which is an aid to inventory control. The main disadvantage is that the contents of ampoules cannot be cooled very quickly; however, the optimum cooling and warming rates of most parasitic protozoa appear to lie within the range attainable with these containers, so this consideration may not apply. For some parasites where the optimum cooling rate lies above that attainable with ampoules, some workers nevertheless prefer to use ampoules and overcome the reduced survival by using much larger volumes of material to initiate their cultures or to infect their laboratory animal maintenance hosts; this practice should be discouraged, primarily because it may place a selective pressure on the parasite population.

It is convenient first to draw out the neck of a glass ampoule in a flame and then score and break off the end so that the parasite suspension has to be introduced using a syringe and needle. This helps to minimize the risk of contamination of the material and reduces the amount of heat required for flame sealing. For ampoules made of polypropylene with screw caps and a silicone rubber sealing ring (e.g. Nunc cryotubes, Gibco-Europe Ltd (Nunc)) the caps should be screwed on tightly prior to final storage to minimize the chance of liquid nitrogen entry.

During cooling, the temperature fluctuations which accompany the evolution of latent heat of fusion are potentially damaging to living cells. This can be overcome with samples in screw-cap ampoules by carefully "seeding" the sample with a small ice crystal once it has been cooled to just below the freezing point. The crystal is deposited using the tip of a Pasteur pipette containing a sample of the suspending medium frozen in liquid nitrogen. This operation is, however, impossible with flame-sealed glass ampoules or capillaries, and seeding in these containers can sometimes be achieved by sharp mechanical agitation or flicking or, alternatively, holding a small piece of metal precooled in liquid nitrogen against the wall of the container.

C. Cryoprotectants

Cryoprotectants can be added in neat undiluted form to the suspension of parasites, but neat glycerol at 0°C is very viscous and so accurate measurement of small volumes is difficult. If added neat, then a table of the glycerol volumes relating to specific suspension volumes to produce the desired final cryoprotectant concentrations should be constructed. An accurately calibrated 0.25 or 1 ml syringe, or an adjustable automatic pipette, is used to add the required volume of glycerol or other additive to the measured volume of parasite suspension, which is then mixed using a Pasteur pipette to draw and expel the suspension three or four times. The same pipette can then be used to fill the capillary tubes or ampoules.

Dimethyl sulphoxide (Me$_2$SO) is a solid at 0°C and reacts exothermically when added to water, so it should first be made up as a double-strength solution in the suspending medium and added in an equal volume to the parasite suspension. Most workers also prefer to use this method of addition with the other cryoprotective compounds.

The temperature of addition of the cryoprotectant is usually approximately 20°C (room temperature) but sometimes 0°C or 37°C, and relatively strict control should be maintained of the duration of exposure to the additive prior to freezing.

III. CRYOPRESERVATION METHODS

It is not really important what method is used to achieve a specific cooling rate, but it is important to know what cooling rate is produced. Programmable freezing machines provide excellent control of the cooling and warming rates, and are particularly useful when experiments are being conducted with new species or strains of parasite to identify the optimum requirements. Many laboratories now use these machines routinely, but they can be relatively expensive and consume fairly large amounts of liquid nitrogen. However, many improvised inexpensive methods have been devised by workers to achieve slow, relatively well-controlled rates.

A. Slow Cooling

One simple method of achieving a cooling rate of just under 1°C min^{-1} over the range 0°C to −60°C for capillaries is to pack them into a plastic holder (see D, below) in an aluminium tube which in turn is placed in an insulated jacket. This is then placed in a refrigerator set at −70°C or below and left overnight. This method is currently in use by the WHO Collaborating Centre for Trypanosomiasis and Cryopreservation of Protozoa at the London School of Hygiene and Tropical Medicine, and has been described in detail in the book "Techniques with Trypanosomes" by Lumsden *et al.* (1973). The heat capacity of the aluminium tube containing the capillaries is approximately 210 joules and the insulating jacket is made of expanded polystyrene with 25 mm thick walls.

When access to a low-temperature refrigerator is not possible, the capillary holder or ampoule can be surrounded with plasticine to a depth of approximately 10 mm and this plug placed into the neck of a liquid nitrogen dewar vessel. Experimentation is required with this method to determine the precise positioning of the plug and the thickness of its walls in order to give the desired cooling rate. A more sophisticated and relatively cheap variable-rate apparatus operating on the same principle is commercially available (Union Carbide, Biofreezer, model BF6).

Capillaries or ampoules can be insulated in a variety of other ways to give various cooling rates when placed in a refrigerator or plunged into a liquid cryogen. Controlled cooling rates can also be obtained using methanol in a dewar vessel. The methanol is agitated continuously using a mechanical stirrer and the dewar is placed inside a larger dewar containing liquid nitrogen or ethanol-solid CO_2. The cooling rate can be varied by altering the amount of methanol in the inner dewar and by substituting double-silvered or unevacuated dewars. The samples are suspended in the methanol. This apparatus and the cooling rates which can be produced are described for *Leishmania* by Callow and Farrant (1973).

B. Rapid Cooling

Since human pathogens have to be sealed in capillaries or ampoules, the fastest cooling rates are those achieved by immersing the sample container directly in a cold liquid. Liquid nitrogen at its boiling point ($-196°C$) is most used, but here a gas envelope forms around the sample container restricting the rate at which heat can be withdrawn from the sample. With capillary tubes, the fastest cooling rate that can be achieved by direct plunge into liquid nitrogen is around $3000-5000°C$ min^{-1} depending on the size of capillary. The insulating layer of gaseous nitrogen is considerably thinner if the sample is instead plunged into melting point nitrogen ($-210°C$) and this will almost double the cooling rate attainable compared to boiling-point nitrogen.

Melting-point nitrogen can be made in the laboratory by reducing the barometric pressure which causes a depression of the boiling point. Nitrogen boils off faster under this reduced pressure and the remaining liquid is cooled by the loss of latent heat of evaporation. The liquid remains at the lowered temperature during the return to normal pressure. By removing sufficient heat from the nitrogen in this way it can be converted to a white solid at $-210°C$. The simplest method of producing melting-point nitrogen is in a standard freeze-drying apparatus, but this procedure is potentially hazardous and expert advice should be sought before attempting this.

Rapid cooling rates can also be achieved by plunging the sample container into a very cold liquid, such as isopentane, precooled to $-160°C$. Heat loss from the sample is extremely rapid since no insulating gas envelope is formed. For non-pathogenic parasites, the sterility of which is not important, even faster cooling rates can be achieved by plunging small sample volumes spread on glass or mica slivvers into liquid nitrogen or other cold liquid, or by introducing the sample in droplet or spray form into the cold liquid.

C. Stepped Cooling

For laboratories with access to a low-temperature refrigerator whose temperature can be regulated, this is one of the simplest methods of cryopreservation. The refrigerator is preset to the desired temperature, usually around -25 to $-35°C$, and the sample simply placed in the cooling chamber for the required length of time and then rapidly transferred directly to the storage temperature.

For *Entamoeba* a holding period of 20 min at $-25°C$ gave good survival (Farri, 1978), while for *Plasmodium* the optimum holding period was 30 min at $-32°C$ (Wilson *et al.*, 1977). Ampoules containing the parasites were brought rapidly to the intermediate temperature by placing them in a methanol bath in a precooled refrigerator. Cunningham *et al.* (1964) also reported that good survival of *Trypanosoma brucei* could be obtained with a two-step cooling method where the samples in capillary tubes were held at $-30°C$ for between 3 and 5 min.

D. Storage and Documentation

Before setting up a bank of cryopreserved parasitic species, some time should be spent in deciding whether the samples will be stored in glass capillaries or ampoules, or in plastic ampoules and what quantity of these containers is likely to be held at any one time. Other considerations will be the ease of transport of the liquid or vapour-phase nitrogen storage container, the frequency with which samples are to be deposited or removed, the static evaporative loss of nitrogen from the vessel, and the availability of a local liquid nitrogen supply. These and other ergonomic considerations will define the size and type of storage vessel and the method of inventory control—on canes, in goblets, or in trays and racks.

Whichever system is finally used it is imperative that comprehensive up to date records are maintained of the material deposited and of the additions to and removals from the bank. A card or computer file of each sample should include:

(1) Details of its origin (genus, species, and strain of parasite, species of donor host or details of patient, geographical location, date, site of infection in host, method of isolation and name of person making the isolation).

(2) Details of any laboratory maintenance or animal passage (including the animal species, size of inoculum, duration of each infection or passage and date).

(3) Details of the preservation (including person doing the cryo-preservation, date, level of parasitaemia, type and amount of cryoprotectant, container, schedule of cooling, and a note of the preferred warming method to be used).

(4) There should also be a record of any viability assays made with thawed material (e.g. count of motile organisms, morphology, infectivity as percentage development relative to a known standard or time taken to reach patency or 2% parasitaemia or the ID_{63} ml^{-1}, pathogenicity, and date).

(5) If the material was cryopreserved as a large batch of tubes or capillaries then a section should be included for recording when these were thawed, to whom they were issued and the number remaining.

Each glass or plastic ampoule should have a unique identification number or code written clearly and indelibly on the outside and its position in the nitrogen storage vessel entered in the card or computer record for that sample. Labelling of individual capillary tubes is difficult, so these are usually put into holders and the relevant information written on the holder. Capillaries can be stored horizontally in trays or, more commonly, vertically in goblets. For this latter method, capillary holders can conveniently be constructed out of triangular section plastic spines (the sort used for loose leaf binding of documents) (Kimber, pers. comm.). A piece of plastic spine is cut with a small saw so that it is approximately 2 cm longer than the capillaries. Two of the three sides are then shortened to leave a 2 cm long tag or handle on the third side, which is used for labelling.

The tubes are prevented from falling out of the bottom of this holder by heat-fusing a triangular piece of nylon gauze to the base. Each holder of this type will accommodate approximately 40 capillaries and the holders can be arranged within the goblets like segments of an orange.

E. Thawing

Almost without exception, most workers use a relatively rapid warming method—agitation of the sample in a water bath usually set to 37°C, but occasionally 40°C or even 45°C. The warming rate produced will depend on the volume of the sample and the type of the material from which the container is constructed. For a 20 μl sample in a glass capillary tube, the warming rate will be in the region of 8000°C min^{-1}, while for 1 ml in a polypropylene ampoule the rate will be about 200°C min^{-1} when agitated in a 37°C water bath. For samples cooled slowly or at their optimum cooling rate, the warming rate will not greatly affect the level of survival. However, if the cooling rate was fast, then the optimum warming rate will also be fast and survival will be much reduced at slow rates of warming.

While there has been very little experimentation to determine the effect of cooling rate on survival of particular parasites, there has been even less testing of the effect of warming rate. Only with trypanosomes (Cunningham et al., 1964) and Plasmodium (Mutetwa, 1983) have different warming rates been evaluated methodically. It was found that rates between about 8 and 2000°C min^{-1} were optimal for trypanosomes previously cooled at 1°C min^{-1} to −60°C and plunged into liquid nitrogen. For Plasmodium cooled at 3600°C min^{-1}, the optimum warming rate was around 8000°C min^{-1}, and this could be achieved simply by removing the capillary tube from liquid nitrogen and agitating it in a 40°C water bath.

The finding of Whittingham et al. (1979) that mouse embryos cooled slowly (0.3 to 0.57°C min^{-1}) to −80°C survived very much better following slow (20°C min^{-1}) than rapid (500°C min^{-1}) warming, suggests that the levels of survival obtained so far with many species of cryopreserved parasites could be improved if investigations were made into variations of the warming rate.

It cannot be said too frequently that improperly sealed glass tubes or ampoules, into which liquid nitrogen has penetrated during storage, pose a significant hazard during thawing. These will explode and should therefore not be warmed directly in a water bath, but in a beaker or measuring cylinder containing prewarmed water.

If pathogens are liberated by an exploding container, then the organisms will be restricted to the beaker or measuring cylinder which can then be appropriately disinfected.

IV. SURVIVAL ASSAYS

Parasitologists have in general been particularly bad at quantifying the survival of their cryopreserved organisms. This is not a particularly important issue when material is simply being banked according to predetermined optimal schedules. However, since in many instances the optimum parameters for cryopreservation have still to be defined, experimentation with techniques will form at least some part of the operations of most workers. Even in the context of experimentation, many published reports have not given the numbers of parasites used, or the volume and concentration either of the material before cryopreservation or upon inoculation into a recipient host or into culture. Furthermore, many workers have failed to use quantitative viability assays for thawed material and have not related the duration of a "prepatent period" to any controls or to a reference standard.

References

Ainsworth, G. C. (1971). "Ainsworth's Dictionary of Fungi". Sixth edition, Commonwealth Mycological Institute, Kew.

Alexander, M , Daggett, P. M., Gherna, R., Jong, S., Simione, F. and Hatt, H. (1980). "American Type Culture Collection Methods I. Laboratory Manual on Preservation Freezing and Freeze-drying". Rockville, Maryland, American Type Culture Collection.

Alston, J. M. and Brown, J. C. (1958). "Leptospirosis in Man and Animals". p. 303. Churchill-Livingstone, Edinburgh.

Annear, D. I. (1956). *J. Hyg., Camb.* **54**, 487–508.

Annear, D. I. (1958). *Aust. J. Exp. Biol. med. Sci.* **36**, 211–221.

Annear, D. I. (1962). *J. Gen. Microbiol.* **27**, 341–343.

Anon. (1971–81). "Index of Fungi". Commonwealth Mycological Institute, Kew.

Anon. (1982). "Mites". Commonwealth Mycological Institute, Kew.

Asher, A. and Spalding, D. (1982). "Culture Centre of Algae and Protozoa, List of Strains". Institute of Terrestrial Ecology, Natural Environment Research Council, Cambridge.

Ashwood-Smith, M. J. and Grant, E. (1977). *In* "The freezing of mammalian embryos". (K. Elliott and J. Whelan, eds) pp. 251–268, Ciba Foundation Symposium 52. Elsevier, Amsterdam.

Atkinson, R. G. (1954). *Can. J. Bot.* **32**, 673–678.

Balch, W. E., Fox, G. E., Magrum, L. J., Woese, C. R. and Wolfe, R. S. (1979). *Microbiol. Rev.* **43**, 260–290.

Balch, W. E. and Wolfe, R. S. (1976). *Appl. Environ. Microbiol.* **32**, 781–791.

Barnes, E. M. (1969). *In* "Methods in Microbiology". (J. R. Norris and D. W. Ribbons, eds) Vol. 3B, pp. 151–160. Academic Press, New York and London.

Barnes, E. M. and Impey, C. S. (1971). *In* "Isolation of Anaerobes" (D. A. Shapton and R. G. Board, eds), pp. 115–123, SAB Technical Series No. 5. Academic Press, London and New York.

Barnes, E. M. and Impey, C. S. (1974). *J. appl. Bact.* **37**, 393–409.

Barnes, E. M. and Impey, C. S. (1978). *In* "Techniques for the Study of Mixed Populations" (D. W. Lovelock and R. Davies, eds), pp. 89–105, SAB Technical Series No. 11. Academic Press, London and New York.

Barratt, R. W., Johnson, G. B. and Ogata, W. N. (1965). *Genetics, Princeton* **52**, 233–246.

Bassel, J., Contopoulou, R., Mortimer, R. and Fogel, S. (1977). *UK Federation for Culture Collections Newsletter* No. 4, p. 7.

Beerens, H., Schaffner, Y., Guillaume, J. and Castel, M. M. (1963). *Annls Inst. Pasteur Lille* **14**, 5–48.

Belcher, H. and Swale, E. (1982). "Culturing Algae, a guide for schools and colleges". Institute of Terrestrial Ecology, Natural Environment Research Council, Cambridge.

Ben-Amotz, A. and Gilboa, A. (1980a). *Marine Ecology—Progress Series* **2**, 157–161.

Ben-Amotz, A. and Gilboa, A. (1980b). *Marine Ecology—Progress Series* **2**, 221–224.

Bennoun, P. (1972). *Compte rendu hebdomadaire des seances de L'Academie des sciences (Paris)* Series D **275**, 1777–1778.

Berger, L. R. (1970). *In* "Proceedings of the First International Conference on Culture Collections". (H. Iizuka and T. Hasegawa, eds) pp. 265–267, University of Tokyo Press, Tokyo.

Bischoff, H. W. and Bold, H. C. (1963). *Phycological Studies* **4**, University of Texas Publication Number 6318.

Boeswinkel, H. J. (1976). *Trans. Br. mycol. Soc.* **66**, 183–185.

Bollinger, R. O., Mussalam, N. and Stuhlberg, C. S. (1974). *J. Parasit.* **60**, 368–369.

Booth, C. (1971a). *In* "Methods in Microbiology". (C. Booth, ed) Vol. 4, pp. 49–94. Academic Press, New York and London.

Booth, C. (1971b). "The genus *Fusarium*". Commonwealth Mycological Institute, Kew.

Brock, T. D. and O'Dea, K. (1977). *Appl. Environ. Microbiol.* **33**, 254–256.

Bryant, M. P. (1972). *Am. J. clin. Nutr.* **25**, 1324–1328.

Bryant, M. P. and Robinson, I. M. (1961). *J. Dairy Sci.* **44**, 1446–1456.

Buell, C. B. and Weston, W. H. (1947). *Am. J. Bot.* **34**, 555–561.

Butterfield, W., Jong, S. C. and Alexander, M. J. (1974). *Can. J. Microbiol.* **20**, 1665–1673.

Butterfield, W., Jong, S. C. and Alexander, M. J. (1978). *Mycologia* **70**, 1122–1124.

Calcott, P. H. (1978). "Freezing and thawing microbes". Meadowfield Press, Shildon, UK.

Calcott, P. H. and Gargett, A. M. (1981). *Fems Microbiology letters* **10**, 151–155.

Caldwell, D. R. and Bryant, M. P. (1966). *Appl. Microbiol.* **14**, 794–801.

Callow, L. L. and Farrant, J. (1973). *Int. J. Parasit.* **3**, 77–88.

Cameron, R. E. and Blank, G. B. (1950). Space Programs Summary. *California Institute of Technology (Pasadena)* **37**, 174–181.

Carmichael, J. W. (1956). *Mycologia* **48**, 378–381.

Carmichael, J. W. (1962). *Mycologia* **54**, 432–436.

Castellani, A. (1939). *J. trop. Med. Hyg.* **42**, 225–226.

Castellani, A. (1967). *J. trop. Med. Hyg.* **70**, 181–184.

Clark, G. and Dick, M. W. (1974). *Trans. Br. mycol. Soc.* **63**, 611–612.

Coe, A. W. and Clark, S. P. (1966). *Mon. Bull. Minist. Hlth.* **25**, 97–100.

Coggeshall, L. T. (1939). *Proc. Soc. exp. Biol. & Med.* **42**,. 499–501.

Coghlan, Joyce D., Lumsden, W. H. R. and McNeillage, G. J. C. (1967). *J. Hyg. Camb.* **65**, 373–79.

Corbett, L. L. and Parker, D. L. (1976). *Appl. environ. microbiol.* **32**, 777–780.

Cowan, S. T. (1974). "Cowan and Steel's Manual for the Identification of Medical Bacteria". 2nd edn. Cambridge University Press, Cambridge.

Crush, J. R. and Pattison, A. C. (1975). *In* "Endomycorrhizas". (F. E. Sanders, B. Mosse, and P. B. Tinker, eds.) pp. 485–509. Academic Press, New York and London.

Cunningham, M. P. (1973). *Int. J. Parasit.* **3**, 583–587.

Cunningham, M. P., Lumsden, W. H. R. and Webber, W. A. F. (1963). *Expl. Parasit.* **14**, 280–284.

Cunningham, M. P., Van Hoeve, K. and Grainge, E. B. (1964). *East African Trypanosomiasis Research Organization Annual Report, July 1963–December 1964*, 26–30.

Dade, H. A. (1960). *In* "Herb Handbook". pp. 78–83. Commonwealth Mycological Institute, Kew.

Daily, W. A. and McGuire, J. M. (1954). *Butler Univ. bot. Stud.* **11**, 139–143.

Dalgliesh, R. J. and Mellors, L. T. (1974). *Int. J. Parasit.* **4**, 169–172.

Dalgleish, R. J. Swain, A. J. and Mellors, L. T. (1976). *Cryobiology*, **13**, 631–637.

Dietz, A. (1975). *In* "Round Table Conference on Cryogenic Preservation of Cell Cultures". (A. P. Rinfret and A. B. La Salle, eds) pp. 22–36. National Academy of Sciences, Washington DC.

Doory, Y. Al-. (1968). *Mycologia* **60**, 720–723.

Ellinghausen, H. C. and McCullough, W. G. (1965a). *Amer. J. vet. Res.* **26**, 39–44.

Ellinghausen, H. C. and McCullough, W. G. (1965b). *Amer. J. vet. Res.* **26**, 45–51.

Elliott, J. J. (1976). *Trans. Br. mycol. Soc.* **67**, 545–546.

Ellis, J. J. (1979). *Mycologia* **71**, 1072–1075.

Eyles, D. E., Coleman, N. and Cavanaugh, G. J. (1956). *J. Parasit.* **42**, 408–413.

Farri, T. (1978). PhD. Thesis, University of London.

Feltham, R. K. A., Power, A. K., Pell, P. A. and Sneath, P. H. A. (1978). *J. appl. Bact.* **44**, 313–316.

Fennell, D. I. (1960). *Bot. Rev.* **26**, 79–141.

Figueiredo, M. B. (1967) *Biologico* **33**, 9–15.

Figueiredo, M. B. and Pimentel, C. P. V. (1975). *Summa Phytopathologica* **1**, 299–302.

Fry, R. M. (1954). *In* "Biological Applications of Freezing and Drying". (R. J. C. Harris, ed.) pp. 215–252. Academic Press, New York and London.

Fry, R. M. (1966). *In* "Cryobiology". (H. T. Meryman, ed.) pp. 665–696, Academic Press, London and New York.

Fry, R. M. and Greaves, R. I. N. (1951). *J. Hyg., Camb.* **49**, 220–246.

Fuller, B., Morris, G. J., Grout, B., Bernard, A., Farrant, J., Pritchard, H. and McLellan, M. (1982). *Cryobiology*. In press.

Gale, A. W., Schmitt, C. G. and Bromfield, K. R. (1975). *Phytopathology* **65**, 828–829.

Giese, A. C. (1973). "Blepharisma: The biology of a light-sensitive protozoan". Stanford University Press, Stanford.

Gilmour, M. N., Turner, G., Berman, R. G. and Kreuzer, A. K. (1978). *Appl. Environ. Microbiol.* **35**, 84–88.

Goldie-Smith, E. K. (1956). *J. Elisha Mitchell Scient. Soc.* **72**, 158–166.

Gordon, W. L. (1952). *Can. J. Bot.* **30**, 209–251.

Gresshoff, P. M. (1977). *Pl. Sci. Let.* **9**, 23–27.

Grivell, A. R. and Jackson, J. F. (1969). *J. gen. Microbiol.* **58**, 423–425.

Hatano, S., Sadakane, H., Nogayama, J. and Watanabe, T. (1978). *Pl. Cell Physiol., Tokyo* **19**, 917–926.

Hatano, S., Sadakane, H., Tutumi, H. and Watanabe, T. (1976a). *Pl. Cell Physiol., Tokyo* **17**, 451–458.

Hatano, S., Sadakane, H., Tutumi, H. and Watanabe, T. (1976b). *Pl. Cell Physiol., Tokyo* **17**, 643–652.

Heaf, D. D. and Lee, D. (1971). *J. Gen. Microbiol.* **68**, 249–251.

Heckly, R. J. (1961). *Adv. appl. Microbiol.* **3**, 1–76.

Heckly, R. J. (1978). *Adv. appl. Microbiol.* **24**, 1–53.

Hill, L. R. (1981). *In* "Essays in Applied Microbiology". (J. R. Norris and M. H. Richmond, eds) pp. 2/1–2/31. John Wiley & Sons, Chichester.

Hilpert, R., Winter, J., Hammes, W. and Kandler, O. (1981). *Zbl. Bact. Hyg., I. Abt. Orig. C* **3**, 149–160.

Hippe, H. and Tilly, I. (1982). XIIIth Int. Congr. Microbiol., Boston, Mass., USA. Abstr. P 54.

Holdeman, L. V., Cato, E. P. and Moore, W. E. C. (1977). "Anaerobe Laboratory

Manual" 4th Edition. Anaerobe Laboratory, Virginia Polytechnic Institute and State University, Blacksburg, Virginia.

Holm-Hansen, O. (1963). *Physiologia Pl.* **16**, 530–540.

Holm-Hansen, O. (1964). *Can. J. Bot.* **42**, 127–137.

Holm-Hansen, O. (1967). *Cryobiology* **4**, 17–23.

Hubalek, Z. and Kochova-Kratochvilova, A (1978). *Antonie van Leeuwenhoek* **44**, 229–241.

Huber, H., Thomm, M., König, H., Thies, G. and Stetter, K. O. (1982). *Arch. Microbiol.* **132**, 47–50.

Hungate, R. E. (1950). *Bact. Rev.* **14**, 1–49.

Hungate, R. E. (1966). "The Rumen and its Microbes". Academic Press, London and New York.

Hungate, R. E. (1969). *In* "Methods in Microbiology" (J. R. Norris and D. W. Ribbons, eds) Vol. IIIB, pp. 117–132. Academic Press, New York and London.

Hutner, S. H. and Provasoli. (1951). *In* "Biochemistry and Physiology of Protozoa". (S. H. Hutner and A. Lwoff, eds) pp. 127–128. Academic Press, New York and London.

Hwang, S.-W. (1960). *Mycologia* **52**, 527–529.

Hwang, S.-W. (1966). *Appl. Microbiol.* **14**, 784–788.

Hwang, S.-W. (1968). *Mycologia* **60**, 613–621.

Hwang, S. and Horneland, W. (1965). *Cryobiology* **1**, 305–311.

Hwang, S. and Huddock, G. A. (1971). *J. Phycol.* **7**, 300–303.

Hwang, S.-W., Kwolek, W. F. and Haynes, W. C. (1976). *Mycologia* **68**, 377–387.

Ivey, M. H. (1975). *J. Parasit.* **61**, 1101–1103.

Jackson, N. E., Miller, R. H. and Franklin, R. E. (1973). *Soil Biol. Biochem.* **5**, 205–212.

Jarvis, J. D., Wynne, C. D. and Telfer, E. R. (1967). *J. med. Lab. Technol.* **24**, 312–314.

Jeon, K. G. (ed.) (1973). "The Biology of Amoeba". Academic Press, London and New York.

Jones, J. B. and Stadtman, T. C. (1977). *J. Bact.* **130**, 1404–1406.

Jong, S. C. (1978). *In* "The Biology and Cultivation of Edible Mushrooms". pp. 119–135. Academic Press, New York and London.

Joshi, L. M., Wilcoxson, R. D., Gera, S. D. and Chatterjee, S. C. (1974). *Indian J. exp. Microbiol.* **12**, 598–599.

Kandler, O. (1982). *Zbl. Bact. Hyg., I. Abt. Orig. C* **3**, 149–160.

Kilpatrick, R. A., Harmon, D. L., Loergering, W. Q. and Clark, W. A. (1971). *Pl. Dis. Reptr.* **55**, 871–873.

Kirsop, B. (1974) *J. Inst. Brew.* **80**, 565–570.

Kirsop, B. (1978) *In* "Abstracts of the XII International Congress of Microbiology, 1978." p. 39, München.

Kirsop, B. E. and Bousfield, I. J. (eds) (1982). "Report on the National Culture Collections, 1982". United Kingdom Federation for Culture Collections/United Kingdom National Committee of Commonwealth Collections of Microorganisms. [Available from UKFCC, see Appendix II.]

Knight, S. C., Farrant, J. and Morris, G. J. (1972). *Nature (New Biol.)* **239**, 88–89.

Korthof, G. (1932). *Zentbl. Bakt. Parasitkde, I. Abt. Orig.* **125**, 429–434.

Kramer, C. L. and Mix, A. J. (1957). *Trans. Kans. Acad. Sci.* **60**, 58–64.

Kratz, W. A. and Myers, J. (1955). *Am. J. Bot.* **42**, 282–287.

Lapage, S. P. and Redway, K. F. (1974). "Preservation of Bacteria with Notes on other Microorganisms". Public Health Laboratory Service Monograph Series No. 7 (A. T. Willis and C. H. Collins, eds). Her Majesty's Stationery Office, London.

Lapage, S. P., Redway, K. F. and Rudge, R. (1978). *In* "Chemical Rubber Company Handbook of Microbiology" (A. I. Laskin and H. A. Lechevalier, eds) Vol. 2, pp. 743–758. Chemical Rubber Company Press, Florida.

Lapage, S. P., Shelton, J. E. Mitchell, T. G. and MacKenzie, A. R. (1970). *In* "Methods in Microbiology". (J. R. Norris and D. W. Ribbons, eds) Vol. 3A, pp. 135–228. Academic Press, New York and London.

Last, F. T., Price, D., Dye, D. W. and Hay, E. M. (1969). *Trans. Br. mycol. Soc.* **53**, 328–330.

Latham, M. J. and Sharpe, M. E. (1971). *In* "Isolation of Anaerobes" (D. A. Shapton and R. G. Board, eds) SAB Technical Series No. 5 pp. 134–147. Academic Press, London and New York.

Leach, C. M. (1962). *Can. J. Bot.* **40**, 151–161.

Leach. C. M. (1971). *In* "Methods in Microbiology". (C. Booth ed.) Vol 4, pp. 608–664. Academic Press, New York and London.

Leef, J. L. and Gaertner, F. H. (1979). *J. Gen. Microbiol.* **110**, 221–224.

Leef, J. L., Strome, C. P. A. and Beaudoin, R. L. (1979). *Bull. Wld. Hlth. Org.* **57**, (suppl. 1), 87–91.

Levitt, J. (1980). "Responses of Plants to Environmental Stresses. I. Chilling, Freezing and High Temperature Stresses". pp. 23–64. Academic Press, London and New York.

Loegering, W. Q. (1965). *Phytopathology* **55**, 247.

Long, R. A., Woods, J. M. and Schmitt, C. G. (1978). *Pl. Dis. Reptr.* **62**, 479–481.

Lumsden, W. H. R., Herbert, W. J. and McNeillage, G. J. C. (1973). "Techniques with Trypanosomes". Churchill Livingstone, Edinburgh and London.

Luyet, B.J. and Gehenio, P. M. (1954). *J. Protozool.* **1**, (suppl.), 7.

MacDonald, K. D. (1972). *Appl. Microbiol.* **23**, 990–993.

Macy, J. M., Snellen, J. E. and Hungate, R. E. (1972). *Am. J. clin. Nutr.* **25**, 1318–1323.

Marx, D. H. & Daniel, W. J. (1976). *Can. J. Microbiol.* **22**, 338–341.

Mazur, P. (1966). *In* "Cryobiology". (H. T. Meryman, ed.) pp. 213–315. Academic Press, New York and London.

McGann, L. E. and Farrant, J. (1976). *Cryobiology* **13**, 261–268.

McGowan, V. F. and Skerman, M. D. (1982). "World Directory of Collections of Microorganisms". 2nd edn, John Wiley & Sons, Chichester.

McGrath, M. S. and Daggett, P. M. (1977). *Can. J. Bot.* **55**, 1794–1796.

McGrath, M. S., Daggett, P. M. and Dilworth, S. (1978). *J. Phycol.* **14**, 521–525.

Mehtra, S. K., Dayal, H. M. and Agarwal, P. N. (1977). *Indian J. exp. Microbiol.* **15**, 81–82.

Meryman, H. T. (1966). *In* "Cryobiology". (H. T. Meryman, ed.) pp. 609–663. Academic Press, New York and London.

Miles, A. A. and Misra, S. S. (1958). *J. Hyg. Camb.* **38**, 732–749.

Miller, T. L. and Wolin, M. J. (1974). *Appl. Microbiol.* **27**, 985–987.

Mitic, S., Otenhajmer, I. and Damjanovic, V. (1974). *Cryobiology* **11**, 116–120.

Miyata, A. (1975). *Trop. Med., Nagasaki* **17**, 55–64.

Morris, G. J. (1976a). *Archs. Microbiol.* **107**, 57–62.

Morris, G. J. (1976b). *Archs. Microbiol.* **107**, 309–312.

Morris, G. J. (1976c). *J. Gen. Microbiol.* **94**, 395–399.

Morris. G. J. (1978). *Br. Phycol. J.* **13**, 15–24.

Morris. G. J. (1980). *In* "Principles and Practice of Low Temperature Preservation in Medicine and Biology". (M. J. Ashwood-Smith and J. Farrant, eds) pp. 253–283. Pitman Medical Press, Tunbridge Wells.

Morris, G. J. (1981). "Cryobiology". Institute of Terrestrial Ecology, Cambridge.

Morris, G. J. and Canning, C. E. (1978). *J. Gen. Microbiol.* **108**, 27–31.

Morris, G. J. and Clarke, A. (1978). *Archs. Microbiol.* **119**, 153–156.

Morris, G. J., Clarke, A. and Fuller, B. J. (1980). *Cryo-Letters* **1**, 121–128.

Morris, G. J., Coulson, G. and Clarke, A. (1979). *Cryobiology* **16**, 401–410.

Morris, G. J., Coulson, G., Clarke, K. J., Grant, B. W. W. and Clarke, A. (1981). *In* "Effects of Low Temperatures on Biological Membranes". (G. J. Morris and A. Clarke, eds) pp. 285–306, Academic Press, London and New York.

Morris, G. J., Coulson, G. Meyer, M. A., McLellan, M. R., Fuller, B. J., Grout, B. W. W., Pritchard, H. W. and Knight, S. C. (1983). *Cryo-Letters*. In press.

Muggleton, P. W. (1963). *Prog. indust. Microbiol.* **4**, 191–214.

Mutetwa, S. (1983). PhD. Thesis, London University.

Nagel, J. G. and Kunz, L. J. (1972). *Appl. Microbiol.* **23**, 837–838.

Nei, T. (1964). *Cryobiology* **1**, 87–93.

Nichols, H. W. and Bold, H. C. (1965). *J. Phycol.* **1**, 34–38.

Norton, C. C. and Joyner, L. P. (1968). *Res. vet. Sci.* **9**, 598–600.

Norton, C. C., Pout, D. D. and Joyner, L. P. (1968). *Folia Parasit.* **15**, 203–211.

Obara, Y., Yamai, S., Nikkawa, T., Shimoda, Y. and Miyamoto, Y. (1981). *J. clin. Microbiol.* **14**, 61–66.

Odds, F. C. (1976). *UK Federation for Culture Collections Newsletter* No. 3, December, pp. 6–7.

Ogata, W. N. (1962). *Neurospora News Letter* **1**, 13.

Onions, A. H. S. (1971) *In* "Methods in Microbiology". (C. Booth ed.) Vol 4, pp. 113–151. Academic Press, New York and London.

Onions, A. H. S. (1977). *In* "Proceedings of the Second International Conference on Culture Collections". (A. F. Pestana de Castro, E. J. De Silva, V. B. D. Skerman and W. W. Leveritt, eds). University of Queensland, Brisbane.

Osborne, J. A. and Lee, D. (1975). *J. Protozool.* **22**, 233–236.

Otsuka, S. and Manako, K. (1961a). *Jap. J. Bact.* **16**, 814–818.

Otsuka, S. and Manako, K. (1961b). *Jap. J. Microbiol.* **5**, 141–148.

Page, F. C. (1967). *J. Protozool.* **14**, 499–521.

Page, F. C. (1976). "An illustrated Key to Freshwater and Soil Amoeba". Scientific Publication No. 34. Freshwater Biological Association, Ambleside.

Page, F. C. (1981). "The Culture and Use of Free-Living Protozoa in Teaching". Institute of Terrestrial Ecology, Natural Environment Research Council, Cambridge.

Palmer, D. A., Buening, G. M. and Carson, C. A. (1982). *Parasit.* **84**, 567–572.

Pautrizel, R. and Carloz, L. (1952). *C.r. Seanc. Soc. Biol.* **146**, 89.

Pell, P. A. and Sneath, P. H. A. (1983). *J. appl. Bact.* In Press.

Perkins, D. D. (1962). *Can. J. Microbiol.* **8**, 591–594.

Peters, J. and Sypherd, P. S. (1978). *J. Gen. Microbiol.* **105**, 77–82.

Phillips, B. A., Latham, M. J. and Sharpe, M. E. (1975). *J. appl. Bact.* **38**, 319–322.

Plattner, H., Fischer, W. M., Schmitt, W. W. and Bachman, L. (1972). *J. Cell Biol.* **53**, 116–126.

Polyansky, G. I. (1963). *Acta Protozool.* **1**, 166–175.

Prescott, D. M. and James, T. V. (1955). *Exp. Cell Res.* **9**, 256–258.

Prescott, J. M. and Kernkamp, M. F. (1971). *Pl. Dis. Reptr.* **55**, 695–696.
Pringsheim, E. G. (1946). "Pure Cultures of Algae". Cambridge University Press, Cambridge.
Raper, K. B. and Alexander, D. F. (1945). *Mycologia* **37**, 499–525.
Rechcigl, M. Jr. (Ed) (1978). "Chemical Rubber Company Handbook of Nutrition and Food". Vol. 3. Culture Media for Microorganisms and Plants. Section G Diets, Culture Media Food Supplements. Chemical Rubber Company Press, Florida.
Redway, K. F. and Lapage, S. P. (1974). *Cryobiology* **11**, 73–79.
Reinecke, P. & Fokkema, N. J. (1979). *Trans. Br. mycol. Soc.* **72**, 329–331.
Reischer, H. S. (1949). *Mycologia* **41**, 177–179.
Resseler, R., Riel, J. van, and Riel, M. van (1966). *Annls. Soc. belge Méd. trop.* **46**, 213–22.
Rey, L. R. (1977). *In* "Development in Biological Standardisation. International symposium on freeze-drying of biological products" (V. J. Cabusso and R. H. Regamey, eds) Vol. 36, pp. 19–27. S. Krager, Basel.
Rosenblatt, J. E., Fallon, A. and Finegold, S. M. (1973). *Appl. Microbiol.* **25**, 77–85.
Rowe, T. W. G. (1971). *Cryobiology* **8**, 133–172.
Saks, N. M. (1978). *Cryobiology* **15**, 563–568.
Schwarze, P. (1975). *Biochem. Physiol. Pflanz.* **167**, 353–355.
Sharp, E. L. and Smith, F. G. (1952). *Phytopathology* **42**, 263–264.
Shearer, B. L., Zeyen, R. J. and Ooka, J. J. (1974). *Phytopathology* **64**, 163–167.
Simione, F. P. and Daggett, P. M., (1976). *J. Parasit.* **62**, 49.
Simione, F. P., Daggett, P. M., McGrath, M. S. and Alexander, M. T. (1977). *Cryobiology* **14**, 500–502.
Simon, E. M. and Schneller, M. V. (1973). *Cryobiology* **10**, 421–426.
Siva, V., Rae, K., Brand, J. J. and Myers, J. (1977). *Pl. Physiol.* **59**, 965–969.
Skerman, V. B. D. (1973). *In* "Proceedings of the Second International Conference on Culture Collections" (A. F. Pestana de Castro, E. J. DaSilva, V. B. D. Skerman and W. W. Leveritt, eds) pp. 20–40, University of Queensland, Brisbane.
Sleesman, J. P., Larsen, P. O. and Safford, J. (1974). *Pl. Dis. Reptr.* **58**, 334–336.
Smith, D. (1982). *Trans. Br. mycol. Soc.* **79**, 415–421.
Smith, D. (1983a). *Trans. Br. mycol. Soc.* **80**, 360–363.
Smith, D. (1983b). *Trans. Br. mycol. Soc.* **80**, 333–337.
Smith, D. and Onions, A. H. S. (1983). *Trans. Br. mycol. Soc.* **81**, 535–540.
Smith, R. S. (1967). *Mycologia* **59**, 600–609.
Smith, R. S. (1971). *Mycologia* **63**, 1218–1221.
Snyder, W. C. and Hansen, H. N. (1946). *Mycologia* **38**, 455–462.
Souzu, H. (1973) *Cryobiology* **10**, 427–431.
Staffeldt, E. E. and Sharp, E. L. (1954). *Phytopathology* **44**, 213–214.
Stamp, L. (1947). *J. gen. Microbiol.* **1**, 251–265.
Stein, J. R. Ed. (1973). "Handbook of Phycological Methods". Culture Methods and Growth Media. Cambridge University Press, Cambridge.
Stetter, K. O., Thomm, M., Winter, J., Wildgruber, G., Huber, H., Zillig, W., Janecovic, D., König, H., Palm, P. and Wunderl, S. (1981). *Zbl. Bact. Hyg., I. Abt. Orig. C* **2**, 166–178.
Swaroop, S. (1938). *Indian J. med. Res.* **26**, 353–378.
Takano, M., Sude, J. I., Tabahira, O. and Gyozo, T. (1973). *Cryobiology* **10**, 440–444.

Timnick, M. B., Lilly, V. G. and Barnett, H. L. (1951). *Phytopathology* **41**, 327–336.
Tsuru, S. (1973). *Cryobiology* **10**, 445–452.
Tuite, J. (1968). *Mycologia* **60**, 591–594.
Turner, L. H. (1970). *Trans. R. Soc. trop. Med. Hyg.* **64**, 624–46.
Venkataraman, G. V. (1969). "The Cultivation of Algae". Indian Council of Agricultural Research, New Delhi.
von Arx, J. A. and Schipper, M. A. A. (1978). *Adv. appl. Microbiol.* **24**, 215–236.
von Rehberg, R. (1978) *Branntweinwirtschaft* **118**, No. 1, January.
Walker, P. J. (1966). *Lab. Pract.* **15**, 423–426.
Warhurst, D. C. and Wright, S. G. (1979). *Trans. R. Soc. trop. Med. Hyg.* **73**, 601.
Webber, W. A. F., Cunningham, M. P. and Lumsden, W. H. R. (1961). *East African Trypanosomiasis Research Organization Annual Report, January-December 1961*, 10–11.
Webster, J. and Davey, R. A. (1976). *Trans. Br. mycol. Soc.* **67**, 543–544.
Wellman, A. M. (1971). *Cryobiology* **7**, 259–262.
Wellman, A. M. and Stewart, G. G. (1973) *Appl. Microbiol.* **26**, 577–583.
Whittingham, D. G., Wood, M., Farrant, J., Lee, H. and Halsey, J. A. (1979). *J. Reprod. & Fert.* **56**, 11–21.
Whitman, W. B., Ankwanda, E. and Wolfe, R. S. (1982). *J. Bact.* **149**, 852–863.
Whitton, B. A. (1962). *Br. phycol. Bull.* **2**, 177–178.
Wildgruber, G., Thomm, M., König, H., Ober, K., Ricchiuto, T. and Stetter, K. O. (1982). *Arch. Microbiol.* **132**, 31–36.
Williams, D. L. and Calcott, P. H. (1982). *J. gen. Microbiol.* **128**, 215–218.
Wilson, R. J. M., Farrant, J. and Walter, C. A. (1977). *Bull Wld. Hlth. Org.* **55**, 309–315.
Wolff, J. W. (1960). *In* "The Leptospirae and Leptospirosis in Man and Animals." *The Preblem Session Series of the Polish Academy of Sciences*, **19**, 11–15.
Woods, R. (1976) *UK Federation for Culture Collections Newsletter* No. 2, p. 5.
Yamai, S., Obara, Y., Nikkawa, T., Shimoda, Y. and Miyamoto, Y. (1979) *Br. J. Vener. Dis.* **55**, 90–93.
Zehnder, A. J. B. and Wuhrmann, K. (1977). *Arch. Microbiol.* **111**, 199–205.
Zeikus, J. G. (1977). *Bact. Rev.* **41**, 514–541.
Zeikus, J. G. and Henning, D. L. (1975). *Antonie von Leeuwenhoek Microbiol. Serol.* **41**, 543–552.
Zeikus, J. G. and Wolfe, R. S. (1972). *J. Bact.* **109**, 707–713.

Appendix I

Documentation

An efficient documentation system is an essential accompaniment to culture maintenance and the following lists provide examples of the kind of information that should be recorded. Details of information required will vary between laboratories. Data may be stored in notebooks, files, card indexes, punch cards, or in semi- or fully-computerized systems.

A. STRAIN RECORDS

1. Essential

a. laboratory reference number;
b. alternative reference numbers (service collections, other laboratories);
c. name, if known;
d. date of deposit;
e. cultural requirements (medium, temperature, pH, nutritional requirements . . .);
f. maintenance method (subculture at 4-week intervals/freeze-drying using "*mist-desiccans*"/2-stage freezing in 10% glycerol . . .).

2. Desirable

a. use of culture (production strain, assay strain, sterility testing . . .);
b. cultural properties (physiological, morphological, genetical . . .);
c. references;

B. MAINTENANCE RECORDS

1. Sub-culturing information: medium, temperature, *records*: date, cultural characteristics, special notes . . .
2. freeze-drying information: method of growth and cell concentration of inoculum, suspending medium, number of ampoules to be prepared, storage and revival procedures . . .
   records: date of processing, batch number, viable counts, date for reprocessing, number of ampoules in storage . . .
3. liquid nitrogen storage information: method of growth and cell concentration of inoculum, cryoprotectant, freezing rate, number of ampoules/capillaries/straws to be prepared, storage and revival procedures . . .

records: date of freezing, batch number, survival counts, date of reprocessing, number of ampoules/capillaries/straws in storage . . .

c. "IN-HOUSE" RECORDS FOR SUPPLY OF CULTURES

Date, person requesting culture, purpose of request, feedback on culture performance . . .
Example:

Strain no.	Date requested	Requested by	Purpose	Feedback
XYZ83	6.4.83	BHK	pilot plant run 41	normal performance

Appendix II

Culture collections and federations

A. CULTURE COLLECTIONS

1. American Type Culture Collection
   12301 Parklawn Drive
   Rockville
   Maryland 20852
   USA
2. Centraalbureau voor Schimmelcultures
   Oosterstraat 1
   PO Box 273
   3740 AG
   Baarn
   The Netherlands
3. Collection Nationale de Cultures de Microorganismes
   Institut Pasteur
   28 Rue du Docteur Roux
   75724 Paris Cedex 15
   France
4. Culture Centre of Algae and Protozoa
   36 Storey's Way
   Cambridge CB3 0DT
   UK
5. Culture Collection
   Commonwealth Mycological Institute
   Ferry Lane
   Kew TW9 3AF
   UK
6. Culture Collection of the Institute for Fermentation
   Institute for Fermentation
   17–85 Jugo-Hochmachi 2-chome
   Yodogawa-ku
   Osaka
   Japan

7. Czechoslovak Collection of Microorganisms
   J E Purkyne University
   662 43 Brno
   Czechoslovakia
8. Deutsche Sammlung von Mikroorganismen
   Grisebachstrasse 8
   Göttingen 3400
   Federal Republic of Germany
9. Forest Products Research Laboratory
   Princes Risborough
   Aylesbury
   UK
10. National Animal Cell Culture Collection
    PHLS Centre for Applied Microbiology and Research
    Porton Down
    Salisbury SP4 0JG
    UK
11. National Collection of Dairy Organisms
    National Institute for Research in Dairying
    Shinfield
    Reading RG2 9AT
    UK
12. National Collection of Industrial Bacteria
    Torry Research Station
    PO Box 31
    135 Abbey Road
    Aberdeen AB9 8DG
    UK
13. National Collection of Marine Bacteria
    Torry Research Station
    PO Box 31
    135 Abbey Road
    Aberdeen AB9 8DG
    UK
14. National Collection of Pathogenic Fungi
    Mycological Reference Laboratory
    London School of Hygiene and Tropical Medicine
    Keppel Street (Gower Street)
    London WC1E 7HT
    UK
15. National Collection of Plant Pathogenic Bacteria
    Ministry of Agriculture, Fisheries and Food
    Hatching Green
    Harpenden
    Herts
    UK

16. National Collection of Type Cultures
    Central Public Health Laboratory
    Colindale Avenue
    London NW9 5HT
    UK
17. National Collection of Yeast Cultures
    Food Research Institute
    Colney Lane
    Norwich NR4 7UA
    UK
18. USSR All Union Collection of Microorganisms
    Institute of Microbiology
    USSR Academy of Sciences
    Profsojuznaja 7
    Moscow B-133
    USSR

Addresses of collections other than those listed above may be obtained from the World Directory of Collections of Cultures of Microorganisms, 2nd edition, 1982, Eds V. F. McGowan and V. B. D. Skerman, World Data Center, University of Queensland. The Directory can be obtained from Thomas Rosswall, Secretary, UNEP/Unesco/ICRO Panel on Microbiology, Swedish University of Agricultural Sciences, S-750 07 Uppsala, Sweden, in book form (US$25) or microfiche (US $15) (1983 prices).

B. FEDERATIONS

1. *UK Federation for Culture Collections*
   Secretary: Dr M Richards
   Beecham Pharmaceuticals Research Division
   Chemotherapeutic Research Centre
   Brockham Park
   Betchworth RH3 7AJ
   UK
2. *World Federation for Culture Collections*
   Secretary: Dr A Onions
   Commonwealth Mycological Institute
   Ferry Lane
   Kew TW9 3AF
   UK

Appendix III

List of Suppliers

Adelphi Manufacturing, 20–21 Duncan Terrace, London, N1 8BZ, UK.

Anchor Glass Co. Ltd, Brent Cross Works, North Circular Road, London, NW4 1JS, UK.

Armour Pharmaceuticals Co. Ltd, Hampton Park, Eastbourne, Sussex, BN21 3YG, UK.

Baird and Tatlock (London) Ltd, P.O. Box 1, Freshwater Road, Chadwell Heath, Romford, Essex RM1 1HA, UK.

B-D Laboratory Products Division, Becton Dickinson UK Ltd, York House, Empire Way, Wembley, Middx. HA9 0PS, UK.

BDH Chemicals Ltd., Poole, Dorset BH12 4NN, UK.

Bellco Glass Inc., Vineland, NJ, USA.

Boots Pure Drug Co. Ltd, Lenton Research Station, Lenton House, Nottingham NG7 2QD, UK.

Rudolph Brand, P.O. Box 310, D-6980 Wertheim, FRG.

Buck & Hickman, Sterling Industrial Estate, Rainham Road South, Dagenham, Essex, RM10 8TA, UK.

Camlab Ltd, Nuffield Road, Cambridge CB4 1TH, UK.

Denley Instruments Ltd, Billingshurst, Sussex, RH14 9SJ, UK.

Difco Laboratories, P.O. Box 14B, Central Avenue, East Molesey, Surrey, KT8 0SE, UK.

Edwards High Vacuum, Manor Royal, Crawley, West Sussex, RH10 2LW, UK.

Creative Beadcraft Ltd., Unit 26 Chiltern Trading Estate, Earl Howe Road, Holmer Green, High Wycombe, Buckinghamshire, UK.

FBG-Trident Ltd., Temple Cloud, Bristol, BS18 5BY, UK.

Genzyme Biochemicals Ltd, Springfield Mill, Maidstone, Kent, ME14 2LE, UK (*formerly* Whatman Laboratory Products Ltd).
   Distributors: Scientific Supplies Co. Ltd, Scientific House, Vine Hill, London EC1R 5EB, UK.

Gibco-Biocult Diagnostics Ltd, 3 Washington Road, Sandyford Industrial Estate, Paisley, PA3 4EP, UK.

Gibco-Europe Ltd., (Nunc), Branch Office, Unit 4, Cowley Mill Industrial Estate, Longbridge Way, Uxbridge, Middlesex, UB8 2YG, UK.

Glass Wholesale Supplies Ltd., 566 Cable Street, London EW1 9EZ, UK.

Harris Manufacturing Co. Inc., Billerico, Mass., USA.

Harshaw Chemicals Ltd., P.O. Box 4, Daventry, Northants, UK.

Instruments de Médecine Vétérinaire, IOBD Clemenceau BP 76, 61300 L'Aigle, France.

International Marketing Corporation, 36 Lenexa Business Center, 9900 Pflumm Road, Lenexa, Kansas 66215, USA.

Jencons (Scientific) Ltd, Cherrycourt Way Industrial Estate, Stanbridge Road, Leighton Buzzard, Bedfordshire LU7 8UA, UK.

R. W. Jennings Ltd, Main Street, East Bridgford, Nottingham, UK.

Koch-Light Laboratories Ltd, 37 Hollands Road, Haverhill, Suffolk CB9 8PU, UK.

Laboratory Environmental Supply Associates, 2 Elmside, Green Lane, Hardwicke, Gloucester GL2 6QF, UK.

London Analytical & Bacteriological Media Ltd, Ford Lane, Salford M6 6PB, UK.

Longs Ltd, Hanworth Trading Estate, Chertsey, Surrey KT16 9LZ, UK.

Mi-Dox Ltd, Smarden, Kent, TN27 8QL, UK.

Millipore (UK) Ltd., Millipore House, 11–15 Peterborough Road, Harrow, Middx., HA1 2YH, UK.

Murphy Chemical Co., Wheathampstead, St. Albans, Herts. AL4 8QY, UK.

Nuclepore Corporation, 7035 Commerce Circle, Pleasanton, CA 94566, USA.

Oxoid Ltd, Wade Road, Basingstoke, Hampshire, RG24 0PW, UK.

Plowden and Thompson Ltd, Dial Glass Works, Stourbridge, West Midlands, DY8 4YN, UK.

John Poulten Ltd, 77–93 Tanner Street, Barking, Essex, IG11 8OD, UK.

Raven Scientific Ltd, P.O. Box 2, Haverhill, Suffolk, CB9 7UU, UK.

Rejafix Ltd., Harlequin Avenue, Great West Road, Brentford, Middlesex, TW8 9EH, UK.

Scientific Supplies Co. Ltd, Scientific House, Vine Hill, London EC1R 5EB, UK.

Scotts Office Equipment Ltd., Deseronto Wharf, St. Mary's Road, Langley, Slough, Berks, SL3 7EW, UK.

Sterilin Instruments, 43–45 Broad Street, Teddington, Middx., TW11 8QZ, UK.
  Distributors: R and L Slaughter Ltd, 162 Balgores Lane, Gidea Park, Romford, Essex, RM2 6BS, UK.

Sweetheart International Ltd, Rowner Road, Gosport, Hampshire, PO13 0PR, UK.

Tissue Culture Services (TCS), 10 Henry Road, Slough, Bucks, SL1 2QI, UK.

Union Carbide, Cryogenics Division, Redworth Way, Aycliffe Industrial Estate, Co. Durham, DL5 6HE, UK.
  Distributors: Jencons (Scientific) Ltd, Cherrycourt Way Industrial Estate, Stanbridge Road, Leighton Buzzard, Beds, LU7 8UA, UK.

Planer Products, Windmill Road, Sunbury-on-Thames, Middlesex, TW16 7HD, UK.

Universal Stationers, Lonsdale House, Empire Way, Wembley, Middx., HA9 0XN, UK.

Wellcome Reagents Ltd, Wellcome Research Laboratories, Beckenham BR3 3BS, UK.

Wesley Coe (Wingent) Ltd, 115–117 Cambridge Road, Milton, Cambridge, CB4 4AY, UK.

Wheaton Scientific Div., Wheaton Industries, Millville, NJ, USA.

Don Whitley Scientific Ltd, Green Lane, Baildon, Shipley, West Yorks, BD17 85JS, UK.

A. D. Wood (London) Ltd, Service House, 1 Lansdowne Road, London N17 0LH, UK.

Genus Index

Subject Index